SAFETY AND RESCUE FOR DIVERS

Also available

Sport Diving – The British Sub-Aqua Club Diving Manual
Advanced Sport Diving
Seamanship For Divers
Snorkelling For All
Teaching Scuba Diving

SAFETY AND RESCUE FOR DIVERS

The British Sub-Aqua Club

EBURY PRESS
London

First published in 1987
Revised edition 1991, 1993, 1998

1 3 5 7 9 10 8 6 4 2

This edition published in 1998 by Ebury Press
Random House, 20 Vauxhall Bridge Road, London SW1V 2SA

Random House Australia (Pty) Limited
20 Alfred Street, Milsons Point, Sydney,
New South Wales 2061, Australia

Random House New Zealand Limited
187 Poland Road, Glenfield, Auckland 10, New Zealand

Random House South Africa (Pty) Limited
Endulini, 5A Jubilee Road, Parktown 2193, South Africa

Random House UK Limited Reg. No. 954009

A CIP catalogue record for this book is available from the British Library.

ISBN 0 09 185311 7

Printed and bound in Scotprint Ltd, Musselburgh, Scotland.

Contents

Foreword

It is an unfortunate fact of life that in all activities accidents will occur. When these accidents occur underwater, the outcome can easily become fatal if the appropriate remedial action is not taken quickly. A significant portion of sport diver training is devoted to teaching would-be divers how to deal with emergency situations. The excellent safety record of the sport attests to the effectiveness of the training given by the recognized sport diving organizations.

This manual provides a wealth of practical information on dealing with diving emergency situations for both the novice and experienced diver alike. Starting with the factors which help to prevent accidents, the manual progresses through the recognition of diving disorders and their primary treatment before progressing to major rescue techniques, life support and first aid.

Throughout the manual, the emphasis is on what can be done by sport divers at the scene of the incident until the casualty can be delivered to qualified medical care. It is not the intent of this manual to give a detailed description of the physiology of the disorders or of the ultimate medical treatment. For such information, the reader is referred to *Sport Diving*, the British Sub-Aqua Club diving manual, to which this manual is a companion volume and to other standard texts.

No manual can do more than describe the techniques that can be used in emergency situations. The reader is therefore advised to use this manual in conjunction with the appropriate practical training provided by the British Sub-Aqua Club, BSAC schools and other recognized sport diver training agencies.

To be able to assist a fellow diver in distress is every diver's responsibility. This manual is intended to help divers fulfil this responsibility.

Trevor Davies
Chief Examiner, Life-Saving Award
British Sub-Aqua Club

Acknowledgements

The British Sub-Aqua Club gratefully acknowledges the efforts of the following persons who have contributed to this publication:

Editors:
Mike Busuttili
Trevor Davies
Mike Holbrook
Gordon Ridley

Contributors
Mike Busuttili
Trevor Davies
Mike Holbrook
Dave Shaw
Cathy Shennan
Dr Peter Wilmshurst
Dave Wybrow

Prevention

Incident Prevention

The Nature of Incidents

Incidents are rarely the result of one causative factor but are more commonly the result of a combination of factors. Each factor in isolation may be quite innocuous, but as the combination builds, the stress on the diver increases ever more rapidly until it is beyond his capabilities.

This effect has been likened to a pit whose sides become steeper and steeper the further one descends. While the slope is gentle, it is easy to retain one's footing and to climb back out. As the descent continues, one's footing becomes less sure, and it becomes more and more difficult to climb out. Ultimately, all grip is lost, and the resulting slide into oblivion at the bottom of the pit is irreversible.

Just as the uncontrollable slide into the pit is avoidable by climbing out before the slope becomes too steep, so can many incidents be avoided if the causative factors are recognized and remedied before they can combine.

Stress and Panic

Stress is a factor that appears whenever a person is placed in an environment or a situation in which they do not feel comfortable. The cause may be physical, such as the effects of cold or exertion, or may be entirely psychological, such as darkness or poor visibility. It is unique however, that stress is a factor common to all diving incidents.

A diver's reaction to stress is a highly individual thing. For some divers, stress will be the spur that they require to accomplish something they would not normally be expected to achieve, while other normally competent divers may have their performance substantially impaired. Taken to extremes, severe stress may well result in panic, a factor, which probably contributes to most underwater fatalities.

Familiarity with and a proven competence at dealing with a situation both reduce the stress imposed on the sport diver. Thorough and progressive training in personal skills and controlling emergency situations can counter the stress imposed by entering an alien environment. An adequate standard of physical fitness helps to reduce the stress imposed by exertion, while suitable equipment counters the stress imposed by the environment.

Stress builds up in an insidious manner, and by its very nature is unlikely to be recognized by the individual affected. The diver concerned may even subconsciously try to disguise the symptoms so as not to appear to have 'chickened out'.

The symptoms of stress vary markedly but those that a diver may notice in his partner include but are not limited to the following:

nervousness
changes in mood
rapid irregular breathing
changes in orientation
erratic and unco-ordinated movements
preoccupation with a trivial problem
problems with buoyancy control
no response to signals
freezing
wide, staring eyes
vigorous treading of water

On the basis that prevention is better than cure, let us look at some of the more common factors that contribute to stress. Many can be anticipated and steps taken to minimize their impact, thus avoiding incidents or at least preventing a minor incident from escalating into a major emergency.

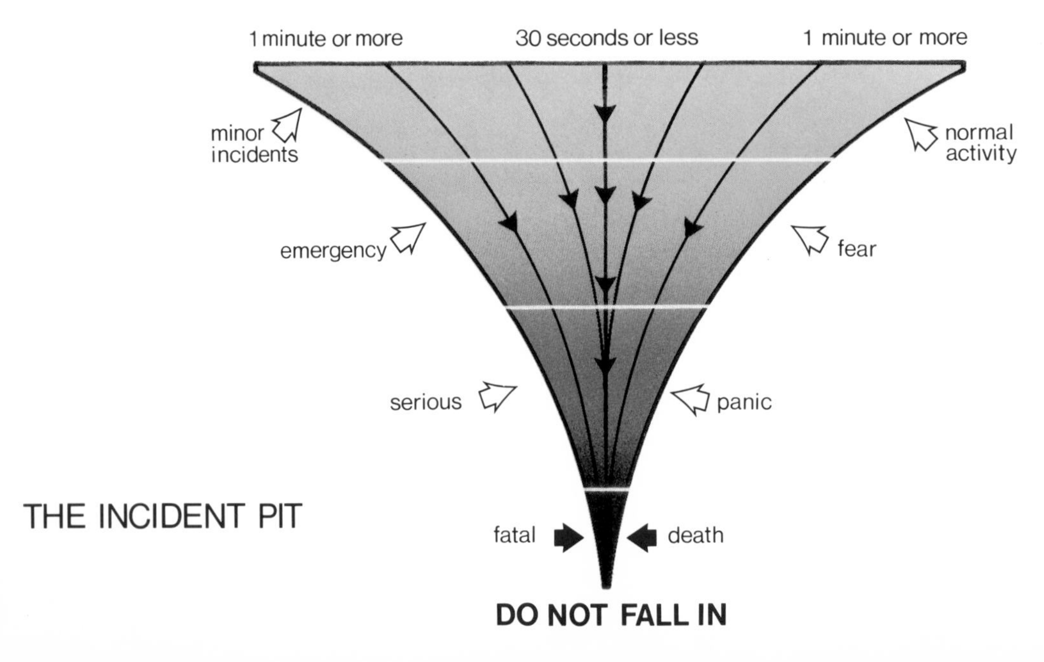

THE INCIDENT PIT

Figure 2

Training

The human being has evolved to function on dry land, breathing air. In order safely to enter the underwater world, a certain amount of adaptation is required. This adaptation consists of learning new skills that will facilitate mobility underwater and learning to use the equipment necessary to permit the human body to function underwater. Additionally, the student diver has to learn how to recognize and deal with emergency situations.

Personal Knowledge and Skills

The underwater environment and the way that the human body reacts to it, impose certain natural limitations on what the diver can do underwater and how long he can remain there. Without a full understanding of these limitations, the diver will be at some risk when venturing underwater. Further limitations will be imposed by the diver's own personal abilities and by the prevailing conditions at the time of diving.

Learning the essential skills of underwater mobility is more than just a question of being able to manoeuvre. The three-dimensional environment and apparent weightlessness are completely alien to human beings, and until the diver masters them, he will not relax underwater. Should some minor incident occur, the added stress of this uneasiness might escalate an easily remedied incident beyond the diver's capabilities.

Use of Equipment

To exist underwater, the diver is dependent on the use of a considerable amount of equipment. While the use of this equipment is straightforward, it is essential that the diver be totally familiar with his own equipment and can operate it reliably. Preoccupation with unfamiliar equipment will detract from the diver's awareness of his surroundings, possibly to the detriment of his or his partner's safety. Routine checking of the equipment should form part of the diver's training in order that impending faults are found before diving commences.

Emergency Skills

Training in emergency skills should not be considered a discrete item, but as an integral part of a diver's training which depends upon competence at, and an understanding of, all other aspects of diver training. A diver who has not been taught to fin properly will have difficulty learning to tow a casualty. He will also have difficulty securing the casualty at the surface if he has not been taught how to operate the type of buoyancy aid that the casualty is wearing.

The real-life emergency situation will always be different from the training situation, and consequently, divers must be taught not only how to do things but also why. This will help them adapt their training to the real situation – should that eventuality occur. Divers must think about what they are trying to achieve, not just mechanically follow a set sequence of actions.

The easier an emergency skill is to perform the more likely it is to be successful. Divers should, therefore, be taught to choose the easiest effective option available to them in an emergency. Training in emergency skills should simulate the real situation as closely, and as safely, as possible. The introduction of 'artificial difficulty' to the training, of the 'if you can cope with this you can cope with anything' variety, is counterproductive. At best this teaches a reaction to a totally non-representative situation and at worst, may pre-condition the introduction of unnecessary difficulty in the real eventuality.

When assessing a diver's performance of emergency skills, the only assessment that has any value is: 'If this was being done for real, would it work?' Pedantic insistence on detail, which does not affect the effectiveness of the technique, is counterproductive. It should also be remembered that standardization between instructors is a useful tool to avoid student confusion but, like any other tool, there are times when its use may not be the optimum method. Standardization should be used for the benefits, which it confers, but should not be allowed to turn into dogma.

Continuity Training

During initial training, divers will become proficient in emergency skills, but the subsequent requirement to use these skills will be very infrequent. The danger in this situation is that when these skills are required in a real emergency, proficiency may have deteriorated to the point where the diver is unable to cope. Tragedies have occurred in the past because divers have no longer been able to cope with what should have been a minor incident. Even more tragic have been the cases where divers have found that the rescue attempt they were making was beyond their capabilities and that, far from being the rescuer, they themselves were in need of rescuing.

Regular practice of emergency skills is necessary to maintain proficiency and ensure that should the necessity ever arise, the diver will be best able to deal with the situation.

Personal skills will also deteriorate if a diver has a break from regular diving due to illness or seasonal weather variations. Although proficiency will soon be regained, this is best done under calm conditions. The progression to more arduous diving can then be made more safely.

Only by continual attention to all the above aspects will a diver maintain confidence in his own abilities and hence be able to relax underwater.

Fitness

By the very nature of the sport, diving requires a reasonable standard of fitness and good health. Because the would-be diver may, completely unknown to himself, have a medical condition which, while of little consequence on dry land, would prohibit him from diving, a diving medical examination is recommended before taking up the sport. Periodic repeat medical examinations are also recommended in case adverse medical conditions develop at a later date.

The fitness for diving of a generally healthy individual may vary from day to day due to the factors discussed below.

Physical Fitness

Under good conditions, once fully submerged and hence effectively weightless, the diver may expend little energy during the course of a dive. Under less than good conditions where the diver may have to cope with a tidal stream or a swell, quite a bit of energy may be expended. Add to this the energy required entering and exiting the water carrying heavy equipment. A good standard of physical fitness is required.

Although not requiring Olympic standards of fitness, a moderate amount of regular exercise will maintain the average diver in good condition. If adequate fitness is not maintained, the diver may become fatigued during a dive and become a potential risk to himself and his buddy. Should the diver's buddy get into trouble, the diver may find himself incapable of effecting a rescue, with dire consequences for the diver in trouble.

Alcohol

Alcohol and the increased pressure underwater have similar effects on the diver. Both impair the diver's mental faculties with the result that not only is the diver's mental performance reduced, but also his ability to realize what is happening is limited. Either factor on its own requires careful monitoring, but the two together are a potentially deadly combination.

The problem can be avoided by not drinking alcohol until after the diving is finished. Even so, a word of caution is required as it takes a long time for alcohol to be eliminated from the blood circulation. After a heavy night's drinking, a diver will still have a significant amount of alcohol in his body the following morning. This is even more insidious as, refreshed by a night's sleep, the diver will be less likely to appreciate the effects of the alcohol.

In addition to dulling the senses, alcohol dilates the peripheral blood circulation. This counteracts the body's natural ability to prevent heat loss. The result is that the diver gets cold more quickly, and an uncomfortable, and in the extreme, a hazardous dive results.

Figure 3 Excess alcohol should be avoided before diving

Figure 4 Smoking is detrimental to divers' health

Drugs

Many drugs used as medications for minor ailments produce side effects. These are often minor, but all are well known to the medical profession, who take them into account when prescribing medicines. Under increased pressure, however, the effects of many drugs change considerably. Some increase the side effects, some decrease them, and others show entirely different side effects. The vast amount of controlled investigation that would be necessary means that very few drugs can be considered to be safe for use under pressure. Many drugs are known to be unsafe. The opinion of a diving doctor should be sought before diving while taking any drugs.

In particular, it is dangerous to dive while taking any drugs, which carry a warning against driving cars, operating machinery, or to the effect that they may cause drowsiness.

Smoking

The detrimental effects on health of smoking have been well publicized and the normal discussion on this topic does not need repeating here. There are, however, additional considerations for the smoking diver to be aware of.

The inhalation of smoke interferes with the normal elimination of respiratory tract secretions. In the lungs, irritants in tobacco smoke may weaken the structure of the lungs, while the increase in mucus may cause blocking of some of the small airways. Both these effects will increase the risk of air embolism and evidence indicates that this is, in fact, more prevalent in smokers than non-smokers.

The carbon monoxide present in tobacco smoke combines readily with the haemoglobin in the blood, thus reducing the blood's capacity to carry oxygen. As it takes many hours for the body to eliminate carbon monoxide, the smoking diver's circulation will be functioning at a reduced oxygen-carrying capacity throughout the dive. In fact, the exhalations of a moderate smoker will contain a far higher proportion of carbon monoxide than would be acceptable in the air in his diving cylinder! The reduced oxygen-carrying capacity of the blood will probably go unnoticed until the diver is required to exert himself, possibly in an emergency, when the diver may quickly become short of breath and may be unable to cope.

Diet

A sensible diet is an important part of keeping fit. Over-indulgence not only reduces fitness, but the accumulation of fat is of particular relevance to the diver. Nitrogen dissolves in fat far more readily than it permeates muscle and tissue. Overweight divers run an increased risk of decompression illness.

Large meals should be avoided before diving as the ensuing conflicting requirements placed on the circulation to satisfy both the digestive system and the muscles will result in a bodily compromise – cramp! Needless to say, anyone who suffers from motion sickness should avoid greasy food before venturing out on the sea.

Motion Sickness

For sufferers of motion sickness the sea can be a miserable place. A number of measures can, however, be taken that will help alleviate the situation and possibly enable the diver to dive safely.

A range of drugs is available to help control motion sickness, although many produce drowsiness as a side effect. These drugs should be tried before going diving to check for side effects, and alternatives tried until one is found that is safe. Generally speaking, anti-motion sickness pills that contain antihistamines should be avoided.

Once on the boat, remain towards the stern where the amount of movement will be least. Keep warm and avoid enclosed spaces and the engine exhaust. Spend as little time as possible fully kitted in heavy equipment before entering the water.

If these measures are ineffective, do not persist in diving. Once in the water the motion will be quite different and may provide a measure of relief. This should not be relied upon for anything but the mildest queasiness. More severe symptoms are unlikely to subside quickly and a diver suffering from motion sickness is at least partially, if not wholly, preoccupied by his condition. With less than his full concentration on the diving in hand this diver is a potential hazard to himself and his buddy. In the extreme case, a diver could submerge only to be sick at depth and vomit into his demand valve, thus putting his life at severe risk.

Figure 5 Motion sickness can incapacitate a diver

Equipment

A diver's life literally depends upon the equipment that he is using. It is therefore essential that all the equipment used is reliable and in good working order.

As the equipment is assembled and put on before diving, each item should be checked for damage and, where possible, checked for correct functioning. After each use the equipment should be washed in clean water, thoroughly dried and then carefully stored until it is required again. At all times the equipment should be treated with respect.

Each year the regulator should be serviced by a recognized service agent and cylinders visually inspected internally. Cylinders should also be hydraulically tested four years from new with a visual inspection every two years. All work should be carried out by properly trained individuals. Amateur adjustments and modifications are potential liabilities.

There is a vast and expanding range of equipment on the market, not all of which is mutually compatible. Technical incompatibilities may exist, such as where the working pressures of various items may differ. Incompatibilities may also be due to one item of equipment, when worn, obscuring access to the controls of another, such as BCs that prevent access to drysuit controls. The diver should regard all the equipment worn not as a set of individual items but as a system, and should keep the operation of the total system in mind when acquiring equipment.

One item of equipment that it is strongly recommended all divers wear, irrespective of their standard of protective clothing, is a buoyancy compensator (BC). While designs may differ in appearance, the essential characteristics of a BC are that it will support a completely equipped diver on the surface and can be fully and rapidly inflated by a means independent of the diver's air supply.

Air Supply

The quality and continuity of a diver's air supply underwater is vital. Cylinders should only be filled from compressors which are being operated correctly and conscientiously. Even though checked when filled, the cylinder pressure should be checked again immediately prior to diving in case a leaking cylinder valve has allowed some or all of the air to escape.

Throughout the dive, cylinder pressure should be frequently checked. Even when considered to be more than adequate for the dive planned, unexpected effort due perhaps to a slight tidal stream may cause the air to be used up much faster than expected. The dive should be terminated with sufficient air, not just to reach the surface, but with a reserve in case any problems arise during the ascent. The amount of this reserve will be dependent on the cylinders being used and the depth/duration of the dive.

Figure 6 It is vital to check your buddy's equipment prior to diving

The Buddy System

The concept of the buddy system is that no diver dives alone so that, should something untoward happen during the dive, there is always someone to help. The system also allows for a measure of incident prevention by providing a means of cross-checking air consumption, depth, and elapsed time to guard against insufficient monitoring or instrument malfunction. Divers should also monitor their partner's reactions for signs of abnormal behaviour that may indicate that their partner has a problem of which he himself may be unaware e.g. nitrogen narcosis.

Merely diving in pairs will not, however, ensure that the buddy system will work. The most common failure of the system is that, in limited visibility divers become separated and lose contact with each other, thus becoming two solo divers. Once out of sight of each other, neither diver can be an effective back-up to the other.

Even in good visibility the system can fail if the divers stray far enough apart so that, while still being able to see each other it would take too long for one diver to return to the other to be of assistance. Divers must remain not only in sight of each other, but also within assisting range.

Assisting a diver in trouble will most likely require the operation of some of his equipment, such as BC or drysuit controls. Because of the vast range of available diving equipment, some of which is operated in quite different ways from other equipment of similar appearance, divers should familiarize themselves with each other's equipment before diving. At depth in an emergency is not the time to try to figure out how to operate your partner's equipment.

Buoyancy

With modern diving equipment, it is possible for the diver to adjust his buoyancy so that he remains neutrally buoyant throughout the dive. Buoyancy control is a tool for the diver to use to make diving more enjoyable, but like any tool its use must be properly taught. It is essential that the diver fully understands the principles of buoyancy and is completely familiar with the controls of his buoyancy equipment. Buoyancy adjustment can be achieved by altering the weight on the diver's belt, or by the controlled inflation and deflation of a BC or drysuit. Reliance on the latter to avoid having to change the weights on a belt, when for instance diving in fresh water instead of seawater, is a potentially dangerous practice that has resulted in a number of incidents. Similarly, over-weighting in the mistaken belief that a well-inflated drysuit will keep the diver warmer has caused problems.

The aim of buoyancy control is that the diver remains neutrally buoyant at all stages of the dive by wearing the minimum amount of weight and the minimum inflation of his buoyancy device. This avoids the exertion required to counteract the effects of incorrect buoyancy and minimizes the restriction and drag caused by excess weight and an overinflated BC or drysuit.

For a wetsuited diver, buoyancy compensation will be effected by the inflation of his BC, whereas the dry-suited diver has the option of either his BC or his suit. Control of buoyancy will be easier and more reliable if only one volume of air needs to be controlled. As the drysuit will require some inflation to counteract suit squeeze, this volume can also be used to achieve buoyancy compensation. While the drysuit can replace the BC for routine buoyancy compensation, it cannot, due to its inherent characteristics, replace the BC in its role in an emergency. It is for this reason that it is strongly recommended that a BC is worn even with a drysuit.

If neutral buoyancy is achieved as described above, the diver will not waste energy, will have a more comfortable dive and will remain better able to cope with the effort required should an incident occur.

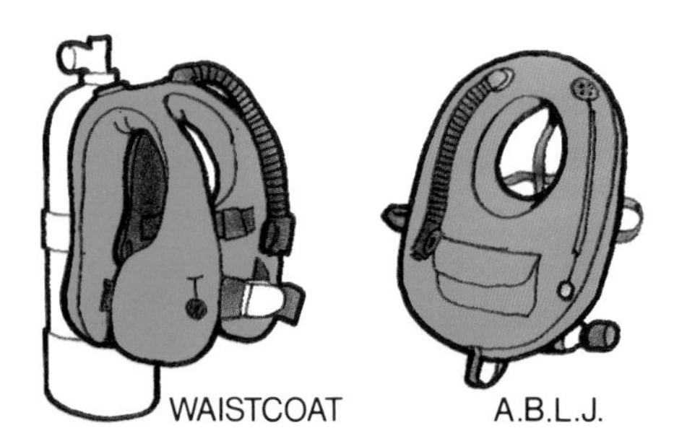

Figure 7 Buoyancy devices

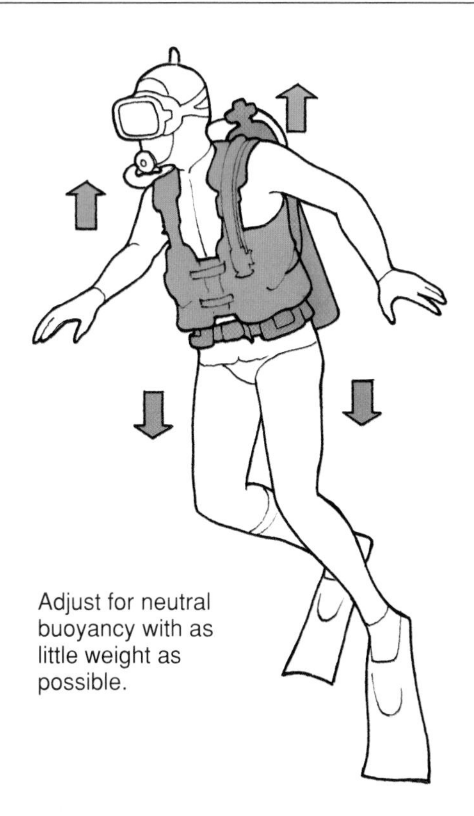

Figure 8

Deep Diving

Deep diving imposes certain physiological limitations on a diver that are avoided at shallower depths. Nitrogen narcosis is encountered which reduces the diver's ability for rational thought, making him prone to mistakes and reducing his ability to respond to any incident. These effects are most severe in the inexperienced or those who have had a lay-off from diving. Even experienced divers are not immune and incidents continue to occur due to divers overestimating their abilities to cope with nitrogen narcosis

Deep diving should be approached by a series of dives of progressively increasing depths (not on the same day), with less experienced divers being partnered by divers with more experience at the planned depth. This will allow the diver to monitor his reactions and to learn to recognize the symptoms of nitrogen narcosis in himself.

The greater absorption of nitrogen during deep diving requires that the dive plan takes into account decompression requirements. The consequences of inadequate decompression can be drastic in the extreme, and meticulous planning is required if this type of diving is to be carried out safely. Because of the attendant risk of nitrogen narcosis, this planning should be carried out on the surface before the dive and the plan religiously followed.

Anticipating Incidents

Anticipating potential factors, which may influence diving safety, is an important element of dive planning. This is equally true for the dive marshal and for the individual divers.

Identification of potential sources of trouble will allow the dive plan to avoid them, or to minimize their influence.

One of the most significant factors that affect dive safety is matching the abilities of the divers to the planned dive. Many incidents occur because divers are allowed to dive in conditions, which are beyond their capabilities. For some dives, therefore, it may be necessary to restrict the diving party to experienced divers. Even then, if the conditions on site are less than optimum, it may be necessary to further restrict the diving party or even to abandon the proposed dive and to dive at an alternative site.

A further extension of this consideration is the makeup of buddy pairs. Less experienced divers will only broaden their experience in safety if accompanied by a more experienced diver. The lack of experienced divers is no excuse for allowing two inexperienced divers to dive together in conditions, which are outside both their experiences.

Having established the diving party, the conditions on site should be kept under review. Dives can be timed to avoid strong tidal streams where a specific object is to be dived or can be deliberately organized as drift dives. The weather is less predictable than the tides. Because the effects of wind and tide combine to affect the sea state the diving conditions may change markedly throughout the course of the day. This should be continuously monitored and, if necessary, the diving curtailed or moved to an alternative location.

Diving safety is considerably improved if the divers concerned and their surface or shore support party all understand the proposed dive plan. The divers should stick to the plan as with limited communication underwater confusion can easily result.

Where diving is taking place from boats it is essential that the boat crew know the location of the divers when in the water, to avoid the possibility of divers being accidentally run down, and to respond quickly should the divers require assistance. When diving on a well-defined site, such as a wreck, this may be achieved by marking the site with a shot line and buoy. Where the site is less well defined, or where divers are drifting with the tidal stream, some other means of marking the divers' position will be needed. Flags on collapsible poles or delayed surface marker buoys carried by the divers will allow them to be seen at the surface from much greater distances, although it should be remembered that this is only effective if the surface party know roughly where to look. A constant indication of precisely where the divers are is given if each pair is equipped with a permanent surface marker buoy.

Should conditions on site deteriorate rapidly or some emergency occurs, it is essential that the boat is able to recall its divers. If surface marker buoys mark the divers the buoy line may be used to signal to the divers using the normal system of rope signals. Where this is not possible, recourse must be made to a diver recall signal. These may be of the explosive 'Thunder-flash' variety or the more recent blank-cartridge-operated variety. They will only be effective, however, if the divers recognize the sound. Experience has shown that divers who have never heard a recall signal before do not recognize it as such and consequently ignore it.

The dive marshal should not only organize the actual diving, but should also ensure that the appropriate emergency equipment is available on site. Depending upon the type of diving being carried out this may be nothing more than a first-aid kit and oxygen equipment or may include such items as VHF radios and flares. When diving from small boats the equipment should also include items necessary for the safe functioning of the boat such as leak stoppers, spare plugs and starter cord for the engine.

The most important item of all for every diver concerned is, however, a liberal application of common sense. With a little more conscientious thought many incidents could be avoided as the train of events which would ultimately end in the incident would be stopped before it started.

Note

Greater detail relating to safety techniques in different forms of diving are covered more comprehensively in the BSAC's *Sport Diving* manual.

Figure 9 Five steps to accident prevention

Drugs and Diving

Drugs

Most people think of drugs as substances used by drug addicts to produce effects, which they find pleasant, or prescribed by doctors to treat diseases. Doctors and pharmacists define drugs differently. To them a drug is a chemical that alters bodily function in a pharmacological way. For an effect to be considered pharmacological there are some important criteria.

Firstly, the substance must produce a dose-response effect. That is, the larger the dose one receives, the greater the effect. An example of this would be the effect of alcohol. The more one drinks, the more drunk one becomes. Another example, which all divers should appreciate, is the effect of increasing partial pressures of nitrogen. When breathing air, the deeper one goes the greater the effect of nitrogen narcosis. At depth, nitrogen, which on the surface is considered inert, produces pharmacological effects by altering the electrical conductivity of nerve cells in the brain.

Secondly, drugs interact with other pharmacological substances. Occasionally one drug increases or potentiates the effect of another, e.g., alcohol will increase the sedative effects of sleeping tablets or nitrogen narcosis. Sometimes a drug will reduce or antagonize the effect of another e.g., amphetamines will reduce the sedation due to sleeping tablets. From this you might deduce that amphetamines would also reduce the sedative effects of nitrogen narcosis. This deduction would be wrong. If a small amount of amphetamine is given to someone breathing air at raised pressure, they will behave very strangely. The effects are unlike those of either nitrogen narcosis alone, or amphetamines on the surface. The interaction here results in a paradoxical effect.

The possibility of interaction of drugs with the physical effects experienced when diving is the reason that there are only a limited number of drugs that divers are permitted to take. There are known to be interactions between a number of everyday drugs and the effects of nitrogen at raised partial pressures. This is particularly apparent for a number of drugs which affect mental function. Drugs, which normally have a mild sedative effect at the surface, including some antihistamines used for treating seasickness and small amounts of alcohol, may considerably increase the effects of nitrogen narcosis. To determine whether a drug may be used safely underwater it is necessary to carry out extensive testing. Clearly, this is a formidable and expensive task. When one realizes that there are many thousands of drugs available, it is apparent that it is almost impossible to guarantee that any drug will be safe if taken before a dive.

Because of the expense involved, the drug testing that has been performed under hyperbaric conditions has been for military or commercial purposes and has little relevance to the drugs which a sport diver is likely to take. Thus, the reason that the BSAC permits the use of very few drugs by its members is that either the drug has possible harmful effects for divers, even when on the surface, or that its effects underwater are unknown, although on the surface it may be safe.

An example of a drug, which could have a harmful effect on the surface for divers, would be beta-blocking drugs, which are used to prevent the heart beating too fast. These drugs can have adverse effects by preventing someone increasing their heart rate appropriately on exercise. As a result, a diver taking these tablets, and who had to fin hard against the tidal stream, might find he was unable to do so.

The BSAC are currently discussing the use of this type of drug. Since many people who need beta-blocking drugs have heart disease or high blood pressure they would be banned from diving anyway. However, these types of drugs are also sometimes used for treating other conditions (e.g., migraine) which would not debar the person from diving. In this situation, the diver would be banned from diving on account of the drug, but not necessarily because of the disease. (In practice, someone with migraine on a low dose of beta-blockers might be permitted to dive if he could satisfy a BSAC medical referee that he could increase his heart rate appropriately on exercise).

By the same token most antibiotics are considered to be safe at the depths dived by sport divers. However, the condition for which the drug is taken would affect whether or not diving is permitted. If antibiotics were taken for a urinary infection, diving would be permitted. If the antibiotics were taken because of a chest infection, diving would not be permitted until recovery. In this situation, the antibiotics do not present a risk, but the chest infection poses an unacceptable risk of burst lung.

Although it is usually impossible to say that a drug may be taken safely when diving, it is often quite easy to state that it is unsafe. The reason for this is that, although it requires lots of dives by many divers without side effects being observed to demonstrate safety, if side effects occur after only a few dives and are repeated, then the conclusion is drawn that the drug is unsafe for divers. Unfortunately, many commonly used drugs are known to be unsafe if taken before diving.

Alcohol

Alcohol taken before a dive potentiates the effects of nitrogen narcosis and also increases the possibility of decompression illness and hypothermia. It causes hypothermia by dilating blood vessels so that excessive heat is lost to the surrounding water and at the same time interferes with the metabolism of the body, so that compensatory heat production does not occur. At least part of the increase in the risk of decompression illness, which results from alcohol ingestion, is the result of dehydration. Interestingly, it can be demonstrated that the dehydrating effects of alcohol on the brain persist for many hours, even after the person who has taken it feels that the effects, including the hangover, have entirely resolved. Home Office statistics show that alcohol is associated with 20 per cent of drownings in Britain, and in some other countries the statistics are even more alarming.

Cigarette Smoking

Tobacco smoke contains a number of substances which produce drug effects, including carbon monoxide, nicotine and tar. These affect the function of the cells lining the air passages, so that mucus collects in the lungs, air passages are narrowed and air trapping occurs. There is evidence that burst lung and air embolism are more common in smokers than non-smokers.

Carbon monoxide binds with haemoglobin and consequently reduces the oxygen-carrying capacity of the blood. About 1 per cent of the haemoglobin of non-smokers is bound to carbon monoxide, mainly from motor car exhausts. Smokers of 10–15 cigarettes a day have about 5 per cent of their haemoglobin put out of action by carbon monoxide, whereas smokers of 20–30 cigarettes a day have effectively lost 10 per cent of the oxygen-carrying capacity of their blood. This is equivalent to losing a pint of blood. Clearly, the maximum exercise capacity of heavy smokers will be impaired as a result, quite apart from the effects due to lung damage resulting from smoking.

Carbon monoxide from tobacco smoke has two additional adverse effects. Firstly, it makes the haemoglobin which is carrying oxygen (not carbon monoxide) reluctant to give up its oxygen to the tissues, which increases tissue hypoxia. Secondly, divers who start the dive with a significant amount of their haemoglobin bound to carbon monoxide will be more susceptible than non-smokers to the risk of carbon monoxide poisoning if there is a low level of carbon monoxide contamination of their air supply.

Nicotine can cause spasm or narrowing of air passages as well as the coronary arteries, which supply blood to the heart muscle itself. The narrowing of air passages will increase the difficulty of respiration, which occurs at depth because of the increased density of the air breathed. A raised partial pressure of oxygen can also cause significant narrowing of coronary arteries. This effect may potentiate in the presence of nicotine.

It is neither practical nor desirable to ban all smokers or people who drink alcohol from diving. However, divers should be encouraged not to smoke or drink alcohol prior to diving. Heavy alcohol consumption the previous night has effects that persist the following day.

Other Drugs

It is likely that the prescribed drugs used most frequently by divers are oral contraceptive pills. Although there are theoretical reasons for believing that decompression illness could be more serious because of the possibility of increased blood clotting if a female diver is on the contraceptive pill, there is no proof of this. So no restrictions are placed on women divers using contraceptive pills.

Other drugs commonly used by divers are for nose and ear complaints, especially when nasal congestion interferes with ear clearing. If this is a problem, the diver should think carefully about whether or not he should dive. However, if he decides to dive, preparations that are sprayed up or dropped into the nose are generally safe to use, but preparations, which are taken by mouth, may not be. The reason for this is that the preparations, which are sprayed up the nose, act only at that site and do not get into the blood stream, unless used excessively. Preparations, which are taken by mouth, have to pass into the blood stream to act. As a result they affect organs other than the nasal lining. In particular, these drugs can affect the heart and cause palpitations, or the brain, inducing drowsiness. It is recommended, of course, that a diver suffering from a cold should not dive.

If it is necessary for a diver to take other drugs he should consult a medical referee about the advisability of diving. After stopping medication, the time taken for the effects of the drug to disappear can vary considerably. Generally, the effects of drugs disappear within one week of the last dose. There are, however, a number of exceptions. For divers, the most important group of drugs with long-lasting effects are those used to treat psychiatric illness. Any person who has received such drugs, including anti-depressants or tranquillizers, should not dive until at least three months after cessation of therapy, unless approval to dive earlier has been given by a medical referee.

Clearly, considerably more research is needed into the effects of drugs on diving.

Dive Planning

Good dive planning is a very important aspect of all diving activities for two reasons. Firstly, it ensures that all the activities are carried out in the most efficient manner, and hence provides the maximum enjoyment for all the participants. Secondly, it ensures that the activities are carried out with the maximum safety. It is not the purpose of this manual to detail all the aspects of planning a dive but to discuss those aspects which are directly related to dive safety.

The elements of dive planning can be divided into two phases. The first phase deals with the considerations of the overall activity up to the point of actually diving, and the second phase deals with the individual dive plans.

Overall Planning

At the earliest stages of dive planning decisions have to be made which have a direct bearing on dive safety. Many of these decisions are inter-related and cannot therefore be taken in isolation. For instance, if the purpose is to have a general branch dive, a site will have to be chosen which provides for all grades of diver. This will have to include access to sheltered, relatively shallow water for students, while also giving access to more ambitious diving for the more experienced divers. If, on the other hand, the purpose is to dive on a deep wreck, this will dictate that only experienced divers are involved and that the dive must be carried out under slack water conditions. Hence, the purpose of the dive, the site, the divers involved, and the conditions are all directly interlinked.

While tidal streams are generally predictable, weather and its effect on water surface conditions cannot be predicted much before the actual event. Dive planning should therefore include plans for an alternative dive if conditions on the day turn out to be unsuitable for the intended dive. The lack of an alternative has been known to cause divers to press on with their original plan regardless of the unsuitable conditions, inevitably resulting in incidents, which could have been avoided with a little forethought.

The planning effort should include the assembly of information regarding the emergency services. Even if it is hoped that this information will never need to be used this must not detract from the thoroughness with which it is assembled. Knowing how to summon the emergency services by telephone or radio is in itself useless if the locations of the nearest radio or telephone are not also known.

The diving party may provide certain emergency equipment itself. Such equipment will normally include first-aid kits, oxygen administration equipment and, where a boat is used, flares, radios and all the normal boat safety equipment. This equipment should all be thoroughly checked before the event and any deficiencies rectified. Arrangements then need to be made to ensure that all this equipment is transported to the dive site, usually by allocating this responsibility to members of the diving party.

Individual divers should at this stage thoroughly check all their personal equipment for serviceability and that it is all packed ready. This should include windproof clothing and anything else that may be necessary for diver comfort between or after dives. Arrival on the dive site to find a vital piece of equipment is either missing or

Figure 10 Planning the dive

unserviceable is likely to cause aggravation. Incidents have arisen in the past due to divers trying to overcome this kind of problem by improvisation.

Individual Dive Plans

Plan the Dive

Individual dive planning commences with the division of the diving party, normally into buddy pairs. The abilities of the divers should be matched such that divers inexperienced in the expected diving conditions are not paired together. In conditions of good visibility and with individuals of suitable diving experience, diving groups of more than two, although not ideal, may be considered.

So that there is no confusion, one member of the diving group who has adequate experience of the diving conditions to be encountered should be nominated as the dive leader. He is responsible for the safe conduct of the dive and should make the decisions for the diving group.

The dive plan decided upon should relate to the experience of the divers concerned, the amount of air they have available and the depth of the water. Suitable parameters for terminating the dive should be determined and understood by all concerned beforehand. Depending upon the nature of the dive these parameters may be easily definable; such as a time limit imposed by decompression considerations, or by air consumption reducing the remaining supply to an agreed reserve value. There may, however, be less easily definable reasons for terminating a dive, such as a diver becoming cold or tired, and the briefing should include agreed signals to signify such conditions.

Immediately prior to entering the water each diver should conduct a final check of his own equipment and that of his partner, ensuring that he understands how to operate the controls of his partner's BC and suit and also how to operate the releases of his weightbelt and equipment.

Dive the Plan

Once underwater and with a much reduced capacity for communication, it is essential that the dive plan is followed. Failure to do so will only result in confusion and detract from both the safety of the dive and the divers' enjoyment of it. Changing the dive plan is potentially dangerous, particularly at depths where nitrogen narcosis is significant and a diver's judgement may be impaired.

Throughout the dive, each diver should monitor air consumption, depth and time. Additionally, the dive leader should regularly crosscheck his partner's air consumption and gauge readings. Each diver should also be continually aware of his position relative to his partner; otherwise the protection afforded by the buddy system is lost. Once divers become visually separated, or so far apart that they cannot offer each other timely assistance in the event of an incident, they are effectively lone divers. Underwater, immediate assistance can only be provided by an effective 'buddy system' and it is each diver's responsibility to ensure that it remains effective, not just for his own benefit, but also that of his partner.

At the appropriate time for the termination of the dive the dive leader should clearly give the signal to ascend and should ensure that the divers remain together during the ascent and on the surface until either they are picked up by a cover boat, or they regain the shore. A large proportion of incidents take place on the surface, therefore the dive should not be considered over until all the group are safely onboard a boat or on the shore.

Figure 11 The dive pair are briefed prior to entering the water

Figure 12

First Aid

Assessment

OBJECTIVES

There are three basic objectives of first aid:

1. **To sustain life.**
2. **To limit the effects of the condition.**
3. **To promote recovery.**

First aid only commences treatment; it does not complete it. The casualty must be transferred to a hospital without delay for qualified medical evaluation and treatment.

CASUALTY EXAMINATION

A prompt evaluation of a casualty's injuries and rapid first aid treatment may be life saving.

This section details what a first aider should be looking for and highlights priorities in management.

Priorities
Airways
Breathing
Circulation

This can be regarded as the ABC of resuscitation.

Is the airway clear?
An unconscious person may breathe on his own, provided the airway is clear.

Is the casualty breathing? – Is there an obstruction?
Indicated perhaps by gurgling or noisy respiration.

Does the casualty have a pulse or other indications of a circulation.
Note: Not only the presence of a pulse, but also the rate, whether it is regular, and whether it has a quality – rate, regularity and volume. The quality of the pulse can provide information about the condition of the casualty and whether his condition is deteriorating or improving. A pulse rate over 100 may indicate previously unrecognized haemorrhage and/or shock, particularly if the pulse is thready (not easy to feel). A slow pulse rate under 60 may indicate hypothermia or carotid sinus syndrome. An irregularity in the pulse may indicate a problem with the pumping action of the heart, e.g., myocardial infarction (heart attack).

Figure 13 Is the casualty breathing?

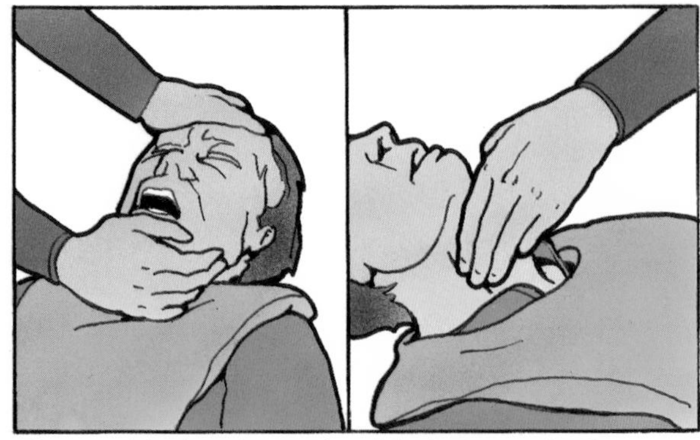

Figure 14 Is the airway clear? Is there a pulse?

Visual Signs

The colour of the casualty can also give vital clues to the first aider. A pale pink colour indicates a good tissue perfusion with oxygenated blood. A pale green indicates seasickness; white, shock; or blue could indicate hypothermia or a respiratory obstruction, or respiratory and/or cardiac arrest.

When dealing with an individual with multiple injuries always follow the ABC of resuscitation. There is no point trying to splint a broken leg if the casualty is not breathing. Having satisfied himself that the casualty is breathing and the circulation is good, the first aider can then turn his attention to wounds, broken bones, burns, shock and other injuries.

Hidden Injuries

Some injuries are obvious, i.e., burns, open wounds or open fractures. In diving, however, other injuries may occur insidiously and be life threatening. For example, cerebral decompression illness. To be effective a diver must acquire evaluation skills of the central nervous system in order to detect the first signs of decompression illness.

The collapse of a diver shortly after surfacing could be due to decompression illness. Therefore, an accurate assessment of his conscious levels should be made.

Does the casualty respond to speech?
If so, is he orientated in time, place and person? i.e., does he know where he is, who he is and what time it is? If a casualty is disorientated on surfacing he should be immediately evacuated to a recompression chamber (trying to differentiate between the different causes of decompression illness in the first aid situation is largely academic as the treatment is the same).

Physical Examination

A physical examination will also assist the first aider in the assessment of the casualty.

Start at the head and work downwards. Look in particular for the following:

Eyes

Can the casualty see?

Have the diver count the number of fingers held up. Try three different numbers. Instruct him to hold his head still. Place your finger in front of the casualty's face and ask him to follow your hand as you move it up and down and to either side. Be sure that both eyes follow in each direction. Look carefully for rapid swinging of the eyeballs. This jerky eye movement is called nystagmus and may indicate brain or vestibular (inner ear) damage.

Face

Ask the casualty to smile!

Check for symmetry. An asymmetrical smile indicates trouble. Use your finger and lightly run it across the casualty's forehead, cheeks and chin whilst he has his eyes closed. Ask him to confirm when you touch him.

Hearing

Ask the casualty to confirm noise.

Rub your thumb and forefinger next to his ear. With his eyes shut, move two-thirds of a metre away and repeat the sound, asking him to confirm it when he hears it. In a noisy area the distance may have to be reduced.

Neck

Ask the casualty to swallow.

Watch the Adam's apple move up and down.

Muscle Strength

Check that the strength appears equal in both arms.

Ask the casualty to stretch both arms straight out in front of him. Check his resistance as you try and push his arms further apart and then further together. In the sitting position have the casualty push out against you with both legs and then instruct him to pull back his legs simultaneously. Check for equal force and reaction in both legs.

Body Sensory Check

Using the same technique on the face, run your finger across his back, shoulders, chest, abdomen, arms and legs. Identify any numb areas.

Balance and Co-ordination

If space permits, have the casualty stand with his feet together and his eyes closed. Ask him to stretch out his hands as far as he can to the sides. Without bending his arms instruct him to bring his index fingers together in one quick movement above his head. Have him alternately touch his nose and your finger held half a metre away as fast as possible.

Any part of this examination, which does not appear to be normal, probably suggests the casualty has decompression illness and should be recompressed immediately.

Note

In all the above examinations remember to take account of pre-existing conditions in the divers.

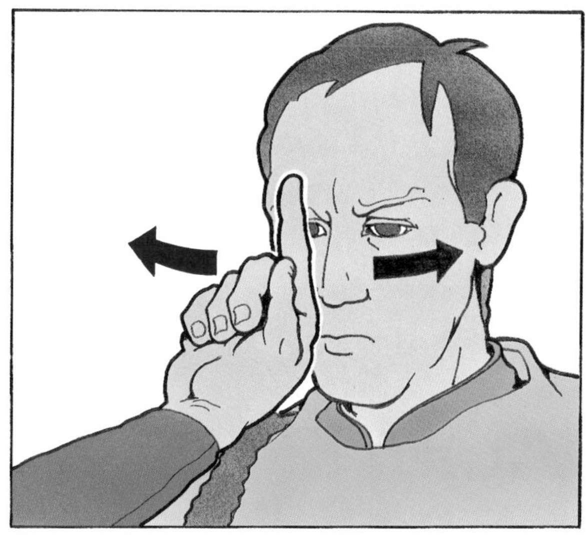

Figure 15 Can the casualty see?

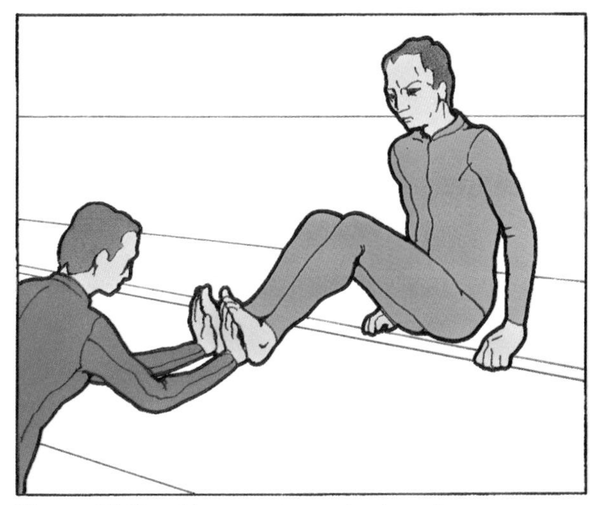

Figure 16 Checking equal muscle strength

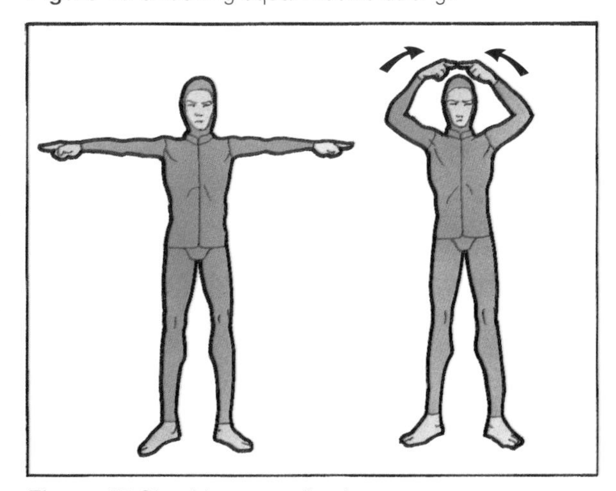

Figure 17 Checking co-ordination

Fractures and Dislocations

Injuries to the bones are rarely a matter of life and death. As a rule, most fractures do not require speed in either treatment or transportation.

A diver whilst kitting up falls over and breaks his leg. He has sustained a simple fracture with no associated soft tissue injury. Well-meaning colleagues rush to help him to his feet. His leg is unable to support his weight and the broken ends of the bone slide past one another, stripping off the covering and tearing the muscles. He falls, only to be picked up again and carried to where he can sit down. With his leg dangling, large blood vessels are torn, and nerves crushed, resulting in paralysis. The jagged ends of the bone come through the soft tissues and finally the skin.

Remember the principles of first aid – to limit the effects of a condition. The manner in which initial care is given to a fracture determines whether the casualty will recover fully in a short time – or suffer a lifetime of disability. Casualties with fractures should be treated and transported slowly and deliberately.

Before considering the first aid treatment of injuries to bone, the different types of injury, their causes and recognition are first considered.

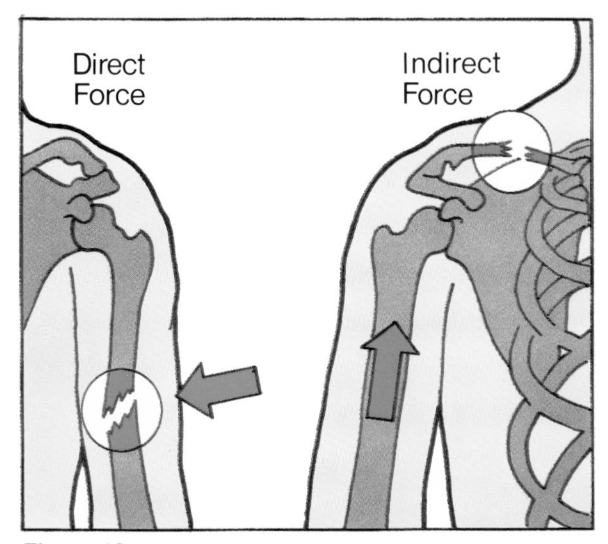

Figure 19

Types of Fracture

For practical first aid purposes, a fracture (a crack or break in bone) falls into two basic categories.

A Closed Fracture

A closed or simple fracture has no associated soft tissue injury. The skin surface is intact.

An Open Fracture

An open wound that extends between the fracture and the skin surface.

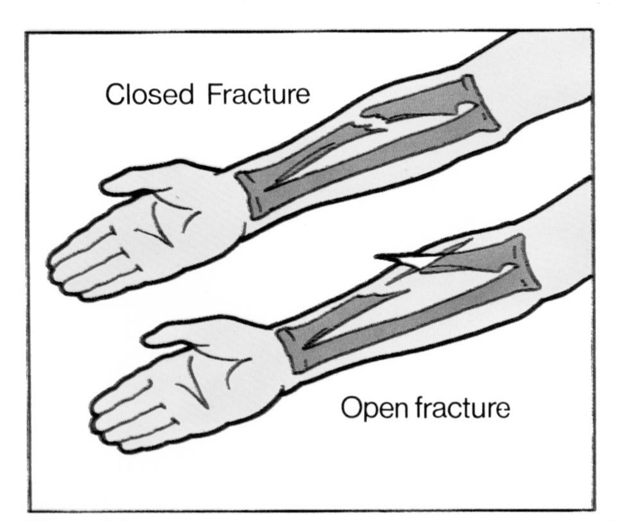

Figure 18

Causes

A fracture may be caused by direct force where the bone breaks at the spot where the force is applied, or indirectly where the bone breaks some distance from the spot where the force is applied. E.g. fracture of the collarbone may result from a fall with an outstretched hand. Here the force is transmitted along the intervening bones, which escape injury (see Figure 19).

Recognition

Exposed bone ends may be visible. Deformity may be apparent. Compare the suspect part with the other arm or leg as appropriate (some people have oddly shaped limbs!).

Depression of the skull would indicate a fractured skull; depression of the ribs would indicate fractured ribs.

The casualty may report feeling the bone break. Pain is made worse by movement of the injured part.

There may be tenderness on gentle pressure over the affected part. There may be crepitus (grating). Here the broken bone ends rub together, but this should never be deliberately sought.

Swelling and discoloration may be present. There may be loss of use of the injured part.

There may be impaired sensation, e.g., numbness may occur if nerve damage is present. A nerve may be pinched or severed by bone fragments. Shock is often caused by concealed haemorrhage from multiple major fractures. For instance thirty minutes after fracture of the femur 25 per cent of the blood volume may be contained in the adult thigh.

Dislocation

A dislocation is the displacement of a bone end that forms part of a joint, e.g., hip, ankle, shoulder or elbow (see Figure 20). The signs and symptoms are similar to that of a fracture with pain, deformity and loss of movement.

Treatment

Casualties with fractures should be transported slowly and deliberately.

Immobilize Before Moving. Splint the injured site before permitting the casualty to be moved (no matter how short the distance) (see Figure 21).. Remember the motto; 'splint them where they lie'. Move the casualty no more than you absolutely have to.

If a fracture is suspected but is not obvious, treat for a fracture. Consider the casualty's total condition, i.e., do not focus only on the fracture and overlook other disorders such as shock.

Do not attempt to reduce a fracture or try to push back any exposed portions of bone.

Expose the fracture site for evidence of skin breakage and dress open fractures before applying a splint, preferably with a sterile dressing.

Remove rings, bracelets, watches or shoes as appropriate from injured area as soon as possible before swelling prohibits their removal.

For fractures of the major bones of the lower limb bring the sound limb gently to the side of the injured one.

A good first aider is a good improviser. Splints can be improvised from paddles or broom handles (see figures 23 and 26).

A good leg can act as a splint for a broken one. If ordinary bandages or triangular bandages are not available for splinting, a long-sleeved woollen jumper makes an excellent alternative for both. It also makes a good arm sling (see Figure 30).

Make sure improvised splints are sufficiently rigid and long enough to immobilize the joint above and below the fracture.

Inspect and carefully prepare a pad before it is applied to make certain the application of the splint does not cause further injury, e.g., pressure damage. Separate skin surfaces with soft padding, e.g., a light woollen jumper, before splinting.

Pass bandages underneath a casualty when he is lying down using the natural hollows of the body, e.g., his knees. Always tie knots over a splint or on the uninjured side (see Figure 21).

A splint should be securely tied, but not so tight that it impedes circulation. Check at fifteen-minute intervals that it is not becoming too tight as a result of swelling of injured tissues. A casualty with numbness because of nerve damage may be oblivious to the fact that tight or poorly padded splints are exacerbating his injury, and a false sense of complacency may occur.

When possible, the injured splinted area should be raised to control haemorrhage and to reduce swelling (see Figure 25).

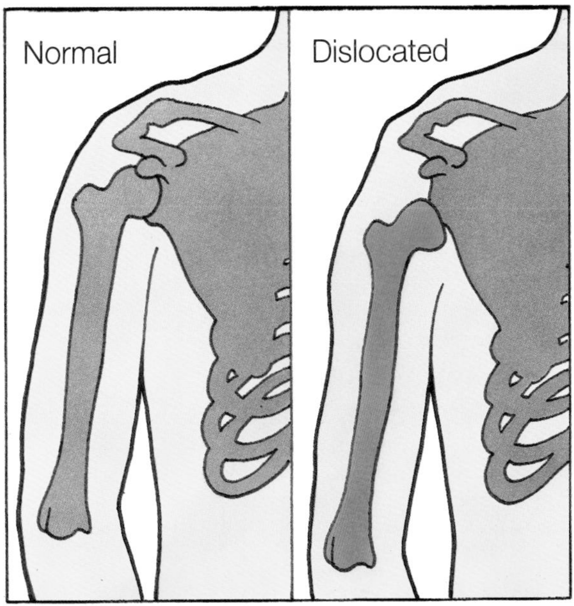

Figure 20 A dislocation

If it is absolutely necessary to move the injured limb, e.g., a broken leg, before a splint can be applied, support the injured extremity both above and below the site of the fracture and have someone apply gentle traction whilst moving, i.e., maintain a slow steady pull on each side of the fracture.

When it is time to move the casualty make certain that persons helping to transport him know what actions to take before starting to move him. Co-ordinated movements are essential. One person should control the injured splinted extremity while the casualty is moved. Do not give the casualty anything to drink or eat. An open fracture will require immediate surgery on admission to hospital. The casualty should be transferred to hospital for further treatment without undue delay.

Splinting a broken bone not only prevents additional damage but also minimizes pain. Unrelieved pain can contribute towards shock. Immediately following a fracture the surrounding muscles are flaccid for ten to forty minutes. The muscle then goes into spasm. Muscular spasms not only interfere with the circulation of blood, but they are painful and increase deformity by pulling bone fragments further out of alignment. Splinting minimizes deformity caused by severe muscle spasm. Splinting also permits blood to clot at the fracture site, preventing further internal bleeding and possible shock.

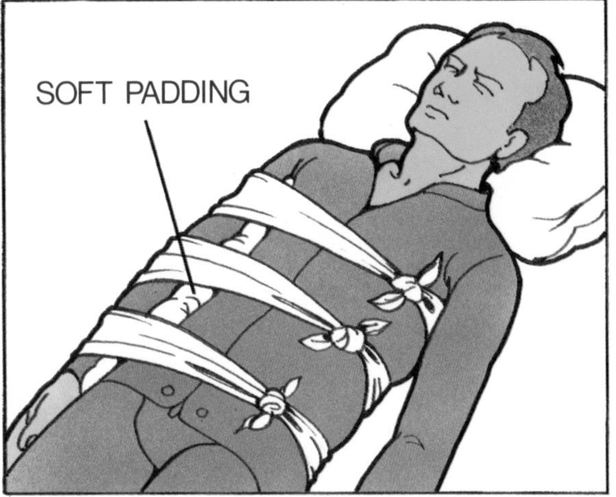

Figure 21 Immobilize before moving

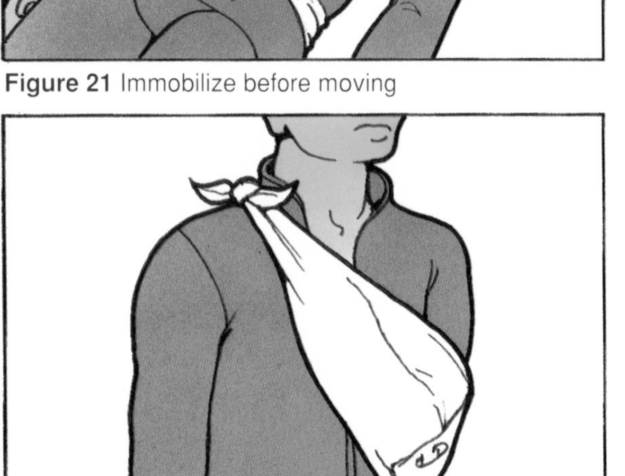

Figure 22 Arm slung from a triangular bandage

Figure 23 Stretcher improvised from paddles and wetsuit

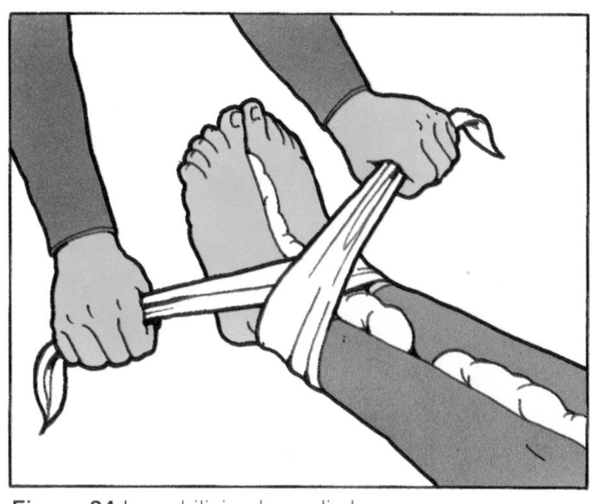

Figure 24 Immobilizing lower limbs

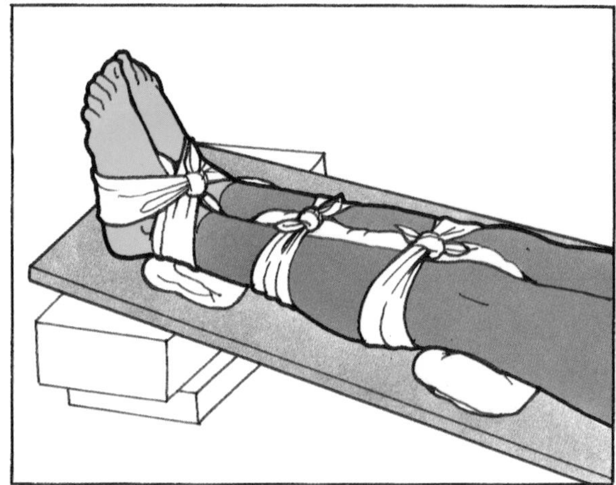

Figure 25 Raise legs to reduce haemorrhage

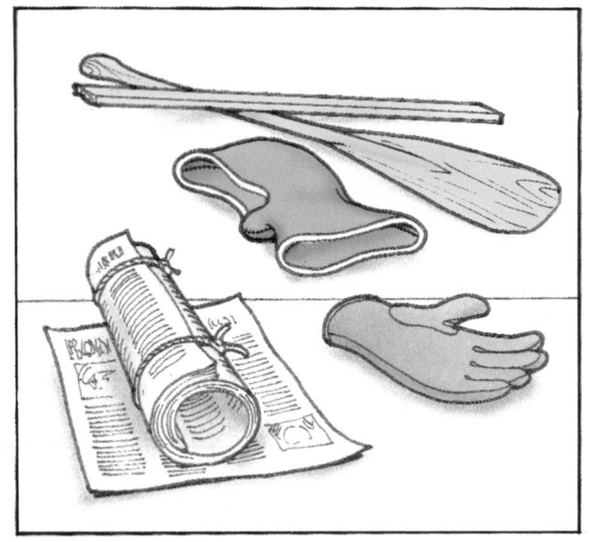

Figure 26 Improvised splints and padding

THE ART OF IMPROVISATION

Cut hand

Once the skin is broken a sterile dressing should be used. However, in an emergency the rescuer may need to improvise.

For a palm injury a torn-up T-shirt will suffice. Made into a pad it can be firmly kept in place by pushing the clenched hand inside a glove. The fingers of the glove are then folded over and strapped in place with a knife strap.

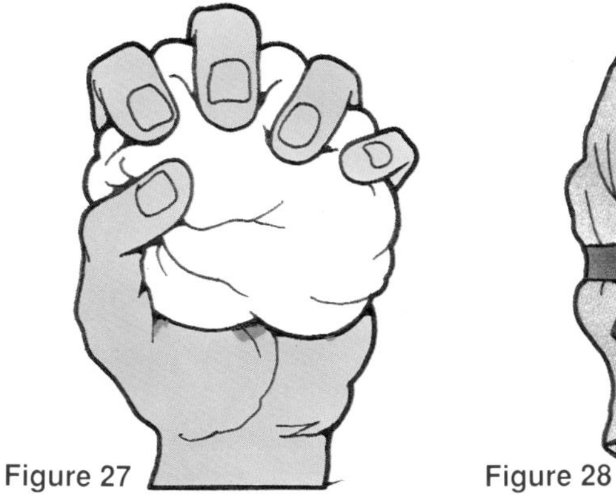

Figure 27

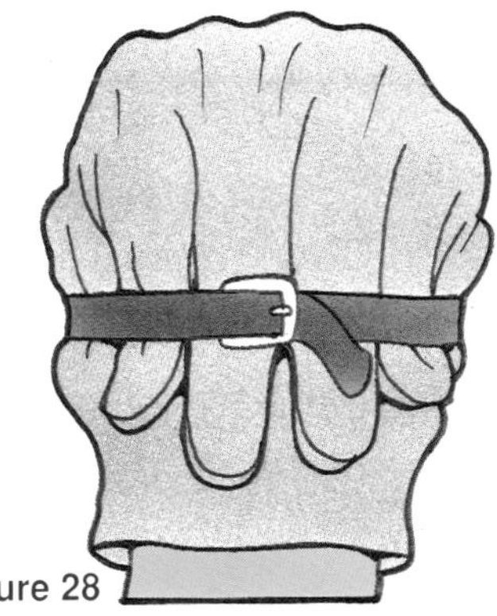

Figure 28

Fracture of the lower limb or thigh

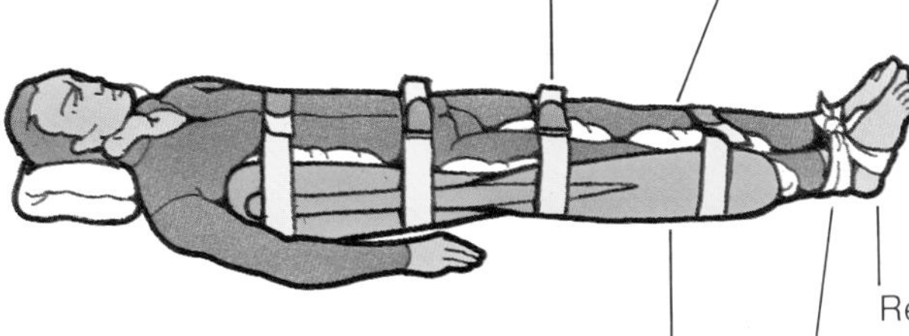

Figure 29

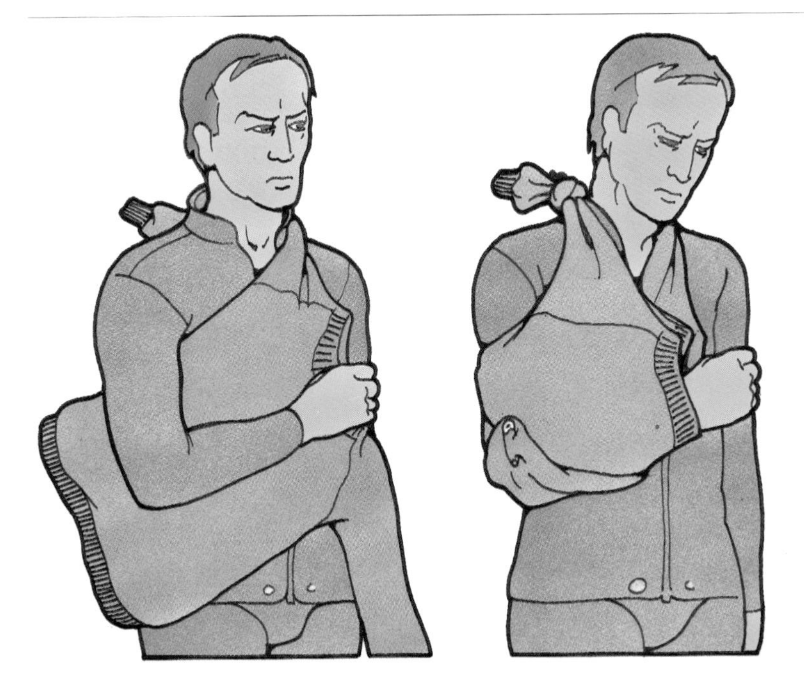

Arm sling

A long-sleeved woollen jumper makes a good triangular bandage; particularly where there is a fracture of the upper limb. It can be used to support the forearm and hand. It is only effective when the casualty is in a sitting or standing position. The hand should be well supported within the sling, but keep the fingers visible for observation. If circulation is impeded the position of the hand should be altered.

Figure 30

Burns

The sort of burns that divers are most likely to come across are the following:

Dry Burns
Dry burns can be caused by flames, lighted cigarettes and hot objects like engines, stoves and kettles which can be found in the galley of hard boats, or caused by fast-moving objects rubbed against the skin. The most common example is a 'rope burn'.

Scalds
Wet heat such as steam, hot water or fat will produce scalds.

Chemical Burns
Acids and alkalis such as those found in cleaning products, may cause burns when they come in contact with the skin or eyes.

Sunburn
The sun's rays can damage the skin or eyes, particularly in an open diving boat in the tropics.

Clothing on Fire
If the clothes of a person become alight they should be extinguished as quickly as possible. Jumping into the water may be the most appropriate way of achieving this, provided it does not place the casualty in danger (consider the boat's propeller). Otherwise the casualty should be put flat on the ground so that the flames do not spread upwards.

Approach the casualty holding a rug, blanket or coat in front of yourself for protection and smother the flames by excluding air (see Figure 31). Do not use nylon. Do not let the casualty run into the open air, as the clothes will burn more than ever because of the increased ventilation and oxygen supply.

Treatment for Minor Burns and Scalds
1. As quickly as possible, immerse the burn in cool water for ten minutes in order to reduce the local effects of heat.
2. Remove anything of a restrictive nature, e.g., rings or boots, before the tissues begin to swell.
3. Cover the burn with a clean, sterile, thick gauze dressing.

Treatment for Severe Burns and Scalds
1. Lay the casualty down and protect the burnt area from contact with the ground.
2. Dowse the area of the burn with copious amounts of cold water.
3. Carefully remove any rings, watches, belts or constricted clothing from the injured area before the tissue starts to swell.
4. Do not remove anything that is sticking to the burn area.
5. Cover the injured area with a thick sterile dressing or similar non-fluffy material and secure with a bandage.
6. Immobilize a badly burned limb.
7. Treat for shock.
8. Reassure the casualty and transfer to hospital.

The casualty with extensive burns may complain of extreme thirst and the temptation to give him too much to drink should be avoided, because this will be followed by intractable vomiting. Only half a glass of water an hour is recommended if there is delay in transfer to hospital.

Do not apply any lotions, ointments or oil dressings. They may be contaminated with bacteria, apart from the fact they may alter the appearance of the burn, making it more difficult for the doctor to diagnose the depth of the burn and the extent of the burned area – knowledge which is essential for correct treatment.

Do not attempt to prick blisters!

Figure 31 Use a blanket to smother flames

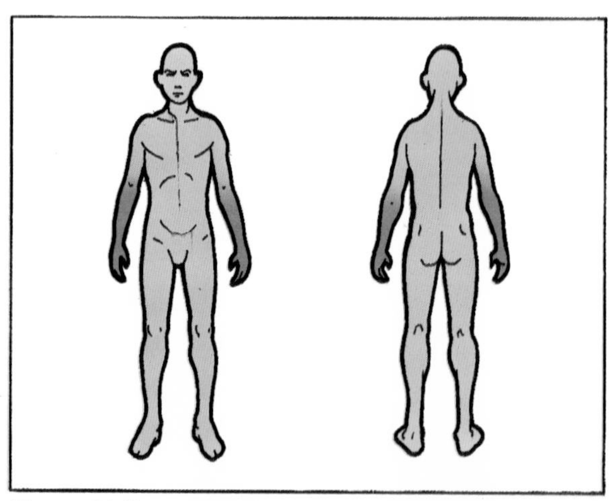

Figure 32 Burns covering 10% of body area

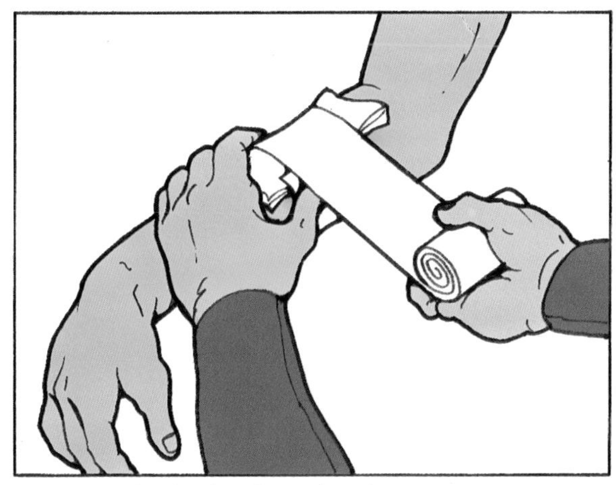

Figure 33 Apply absorbent dressing

The pain of a superficial burn may be very acute. In a superficial burn only the outer layers of the skin are damaged. The casualty should be reassured that the pain will not last long and usually disappears spontaneously. The really deep and most severe burns cause very little if any pain because the whole thickness of the skin, including the nerve endings, is destroyed. The burned casualty who suffers from pain may in fact have a less serious burn.

Treatment for Chemical Burns

1. Flood the affected area with slowly running cold water for at least ten minutes.
2. Carefully remove any contaminated clothing while continuing to flood the injured area.
3. Continue treatment as for severe burns.

Chemical Burns to the Eyes

1. Position the casualty's face under gently running water so that the water drains away from the eye. Alternatively, submerge the eye in a bowl of water and ask the casualty to blink. (Do not allow the casualty to rub the eye).
2. Dress the eye with a sterile eye pad or clean, non-fluffy material.
3. Remove casualty to hospital immediately.

Treatment of Sunburn

Over-exposure to the sun's rays can cause sunburn, especially when a diver's body is wet with seawater. Even when there is a 'refreshing' cool breeze the sun's ultraviolet radiation can cause the skin to become red and blistered.

1. Remove the casualty from direct sunlight, and cool the skin by sponging gently with cold water.
2. Allow the casualty sips of cold water to replace the fluid loss.
3. For severe cases of sunburn, seek medical attention.

Shock

One of the major causes of death in burns is shock. Shock is due to plasma loss from the burned skin in the form of blisters and also into the tissues around the burned area. The plasma leakage is most severe in the first few hours and is greater the larger the area of the body burned. A burn covering more than 10 per cent of the body area is likely to cause shock or possibly death and requires prompt evacuation to a hospital. In the case of scattered burns, it is useful to know that the size of the casualty's hand is about 1 per cent of the area of the body. Any burn exceeding this size should be seen by a doctor.

The classical signs and symptoms of shock may not be present shortly after burning, and may only appear later, when it may be too late to prevent the ill effects of shock.

Infection

The second major cause of death from burns is infection in the form of septicaemia. A burn, by its very nature, is sterile. Try and cover the burn with a 20mm thick sterile gauze dressing held in place with crepe bandages.

The burned area should be completely sealed off from the outside with a thick dressing which serves to absorb any discharge and prevents it reaching the surface, which may result in infection.

Thick dressings are unlikely to be available so improvise with freshly laundered sheets, pillowcases, or a clean handkerchief.

The risk of infection is the reason for not removing the clothes of the casualty.

The casualty with extensive burns often seems remarkably well half an hour after the accident. This apparent well being is deceptive. Changes have occurred, and unless satisfactory treatment is started and continued, these may easily lead to death within the next few hours.

Although the concept of burns shock is widely known and the need for effective treatment is recognized, the urgency of the situation is often overlooked. The insidious leak of plasma from the casualty's circulation is less dramatic than when bright red blood is spurting on the floor, but it can kill just the same.

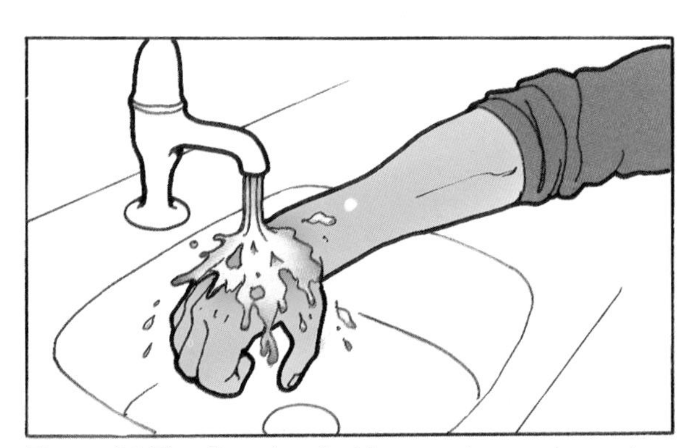

Figure 34 Flooding chemical burn under running water

Bleeding

When faced with a casualty who is bleeding keep calm and take charge. Confidence is infectious (and so is panic!). Ideally, a sterile dressing should be placed over the wound and direct pressure applied (see figures 35 and 36). However, if there are no dressings then improvise. A good first aider is a good improviser. A dry rag from the toolbox is better than nothing. Infection is of secondary importance. If there is nothing immediately available then use your hand.

Tell the casualty to lie still and reassure him. If the casualty is agitated and moving about, he will break up any clot formation that occurs and bleeding will continue. Elevate the affected part – it is more difficult for blood to flow uphill!

If any dressing fails to stop the bleeding do not remove and replace it, as this interferes with the clot formation. Lives have been lost by the removal and replacement of a dressing. Instead, put another dressing on top. If there is a foreign body stuck in the wound do not be tempted to pull it out, as this action may cause massive bleeding.

Major Bleeding

If more than one litre of blood has been lost the casualty will need a replacement transfusion. Speed in transfer to hospital becomes critical.

If bleeding is arterial, direct pressure on the wound may not be sufficient. Bleeding can be controlled here by strong finger pressure on the main artery supplying the affected area. The artery has to be close to the surface of the skin and needs to be compressed against a bone lying underneath it. These areas are called *pressure points* (see Figure 37). Practise identifying pressure point areas by feeling for a pulse before applying pressure. Practise using the brachial pressure point. Apply just enough pressure so that you cannot pick up the radial pulse. Excessive pressure is uncalled for and painful. This method should not be maintained for more than ten minutes.

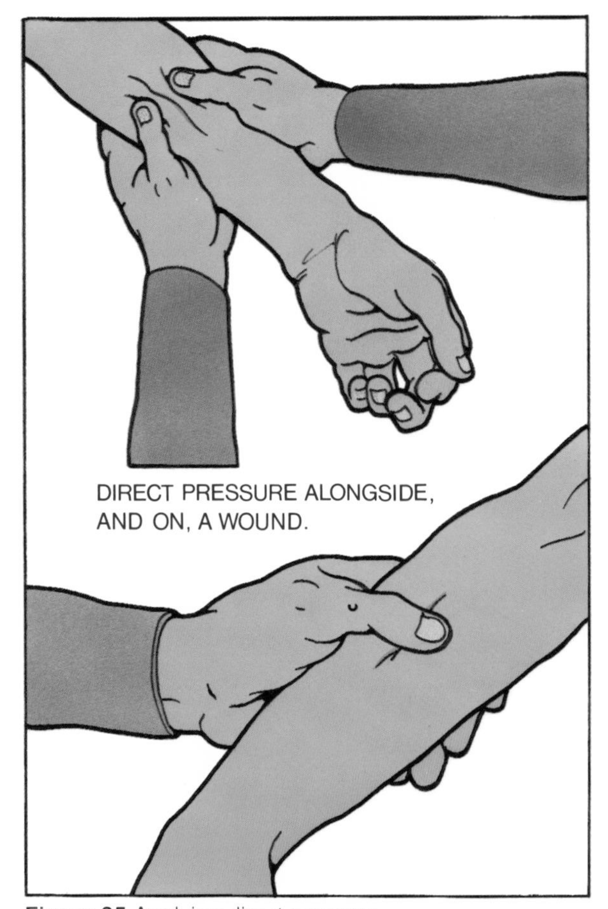

Figure 35 Applying direct pressure

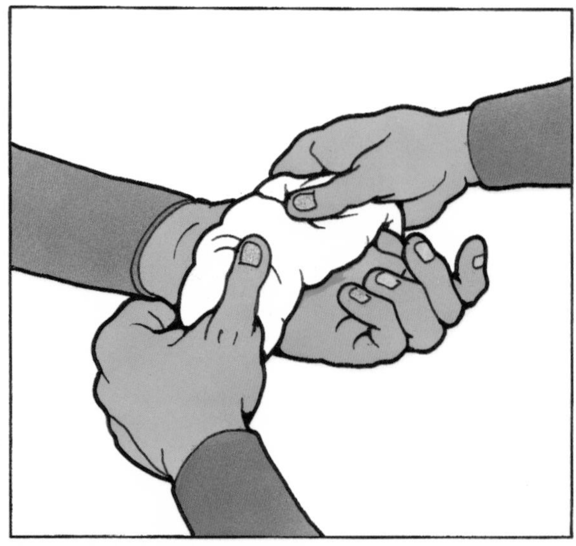

Figure 36 Applying pressure with a pad

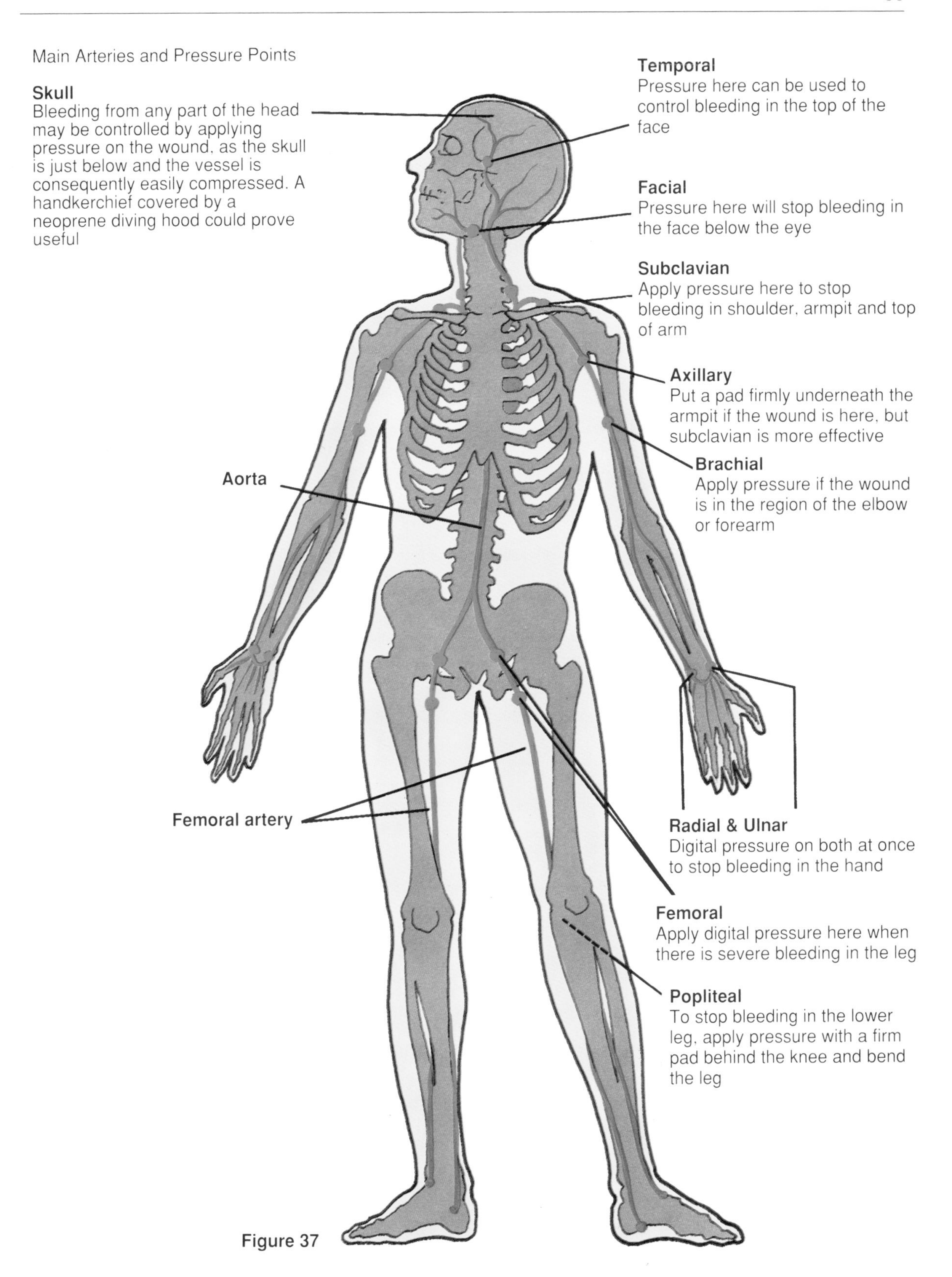
Main Arteries and Pressure Points
Skull
Bleeding from any part of the head may be controlled by applying pressure on the wound, as the skull is just below and the vessel is consequently easily compressed. A handkerchief covered by a neoprene diving hood could prove useful
Temporal
Pressure here can be used to control bleeding in the top of the face
Facial
Pressure here will stop bleeding in the face below the eye
Subclavian
Apply pressure here to stop bleeding in shoulder, armpit and top of arm
Axillary
Put a pad firmly underneath the armpit if the wound is here, but subclavian is more effective
Brachial
Apply pressure if the wound is in the region of the elbow or forearm
Aorta
Femoral artery
Radial & Ulnar
Digital pressure on both at once to stop bleeding in the hand
Femoral
Apply digital pressure here when there is severe bleeding in the leg
Popliteal
To stop bleeding in the lower leg, apply pressure with a firm pad behind the knee and bend the leg

Figure 37

Blast Injuries

If a diving cylinder should rupture for any reason, then this may cause an explosion, with associated injuries resulting from the blast.

There may be other situations in which a diver is likely to encounter blast injuries, for example in the commercial field, where explosives are used in both mining and salvage operations, or among amateur divers exploring military wrecks containing unstable and unexploded ammunition.

Effects of blast injuries on the body

Potential high-velocity fragments from an exploded cylinder in air cause more damage than bullets would as they tear through the tissues in a tumbling fashion. As well as the damage from the fragments, injuries can also be caused by the complex blast wave, which accompanies every explosion. This blast wave consists of a pressure wave and a blast wind.

The pressure wave can have devastating effects, particularly in a confined area. Close to an explosion the pressure wave may rise to 170 bar. The eardrum ruptures at 11.5 bar.

Sheltering behind a wall will not protect you because pressure waves behave like sound waves, flowing over and around an obstruction. Any person standing in front of a wall facing an explosion is subjected to the added effect of reflected pressure.

The blast wind can cause traumatic amputation, or even total body disintegration if sufficiently severe. A person standing behind a wall will be protected from the blast wind, unless of course the wall collapses, causing crushing injuries.

An explosion underwater can have devastating effects on a diver some considerable distance from the explosion, whereas in air, at the same distance, a person would be unharmed. This is because water is an incompressible medium and pressure waves are transmitted with less attenuation. The nature of the injury sustained underwater is considerably different to the injury sustained in air. In air much of the pressure wave is reflected at the body surface because of the different densities of air/body fluids. In water, however, there is little reflection because water and the body fluids have similar densities. When the pressure wave meets an area of different density e.g., an air-filled cavity in the body, considerable damage occurs as the tissues are shredded or torn apart. Areas damaged, therefore, are the ears, sinuses, lungs and gut.

Clinical Features

Casualties from an explosion in air may be injured in a dramatic form with traumatic amputation and severe external bleeding which can be life threatening.

Casualties from an underwater explosion, however, may display no external signs of injury, bruising or laceration despite the profound internal damage. Internal haemorrhage, as a result of this damage, may lead to cardiovascular shock.

Brain damage is thought to be caused by a rapid rise in venous pressure following compression of the chest and abdomen by the pressure wave. The deafness of the victims of blast, due to the rupture of the eardrum, dislocation of the ossicles, or damage to the inner ear, makes communication with them difficult. They must be carefully examined for they may have multiple wounds.

Symptoms may include:

Abdominal or chest pain.
Acute shortness of breath.
Coughing up or vomiting blood *(Haemoptysis/ Haematemesis)*.
Passing black stools or blood from the rectum *(Melaena)*.
Headaches.
Earache.

Management

This is the same as for any other severe body trauma.In addition give 100 per cent oxygen if respiratory distress occurs or oxygen-enriched AV if breathing ceases.
Give nothing to eat or drink.

Remember when transferring the casualty that exposure to altitude may exacerbate any injury, and in particular may precipitate gastrointestinal perforations in damaged areas of the gastrointestinal tract.

All divers exposed to explosions should be admitted to hospital, where they will be maintained on intravenous fluids and gastric suction until the full extent of the damage has been assessed.

Prevention

Ensure that cylinders are tested regularly at a recognized test house. Do not overfill your cylinders, and do not be tempted to do DIY jobs on high-pressure equipment.

Avoid diving where explosions are possible.

Circulatory Shock

Few people seem to have a good understanding of shock. This section defines shock and explores the types of shock that divers are most likely to encounter.

Definition

Shock can be described as 'the inability of the circulatory system to meet the needs of tissues for oxygen and nutrients, and the removal of their waste products'. Because of a lack of adequate tissue blood flow the tissues become damaged and cease to function.

Inadequate tissue blood flow has basically three different causes:

1. The blood volume may be diminished (hypovolaemic shock).
2. The capacity of the circulation is increased by massive dilation of the blood vessels, so that even the normal amount of blood becomes incapable of adequately filling the circulatory system (low-resistance shock).
3. The heart fails to act effectively as a pump, sometimes aptly called the 'power failure syndrome' (cardiogenic shock).

These three causes are now explored in more detail.

Hypovolaemic Shock

Haemorrhage is perhaps the most common cause of hypovolaemic shock, though it will also occur in a case of severe burns where plasma is lost through the exposed areas. Prolonged seasickness can cause continuous vomiting, dehydration and hypovolaemic shock too. It can occur after trauma to the body, even where there is no haemorrhage. A blow, even when the skin is not broken, can often damage the capillaries sufficiently to allow excessive loss of plasma into the tissues. This can result in greatly reduced plasma volume.

Low-resistance Shock

Occasionally, shock results without any loss of blood volume whatsoever. Instead, the capacity of the circulation is increased by massive dilation of the blood vessels. One of the major causes of this is loss of smooth muscle tone in the walls of the arterioles.

The vasomotor centre is responsible for ensuring the smooth muscle in the walls of the arterioles is in a state of partial contraction. Loss of tone causes the arteries to dilate, increasing the diameter of the vessels. An increase of vascular capacity reduces the venous return to the heart, with consequent reduced cardiac output. This effect is often called 'venous pooling' of blood.

An example of this sort of shock is fainting. If a person who has fainted is held in an upright position, he will go into a progressive stage of shock as sufficient oxygen is not reaching the brain, and he can die as a result. Fortunately, on fainting a person usually falls to a horizontal position so that an adequate cerebral blood flow returns almost immediately.

Brain damage is also a cause of vasomotor collapse. Pain sometimes strongly inhibits the vasomotor centre, thereby increasing the vascular capacitance and reducing the venous return.

Cardiogenic Shock

Sometimes the heart itself is damaged so severely that it cannot pump adequate quantities of blood to the tissues; e.g., in a heart attack the cardiac muscle itself cannot contract properly as it is deprived of oxygen. As a result, the tissues deteriorate rapidly and death ensues.

Cardiogenic shock can also be a sequel to pulmonary barotrauma where one lung has collapsed. Because there is now positive pressure on one side of the chest and negative pressure on the other, the heart is shifted towards the unaffected side (mediastinal shift). This causes kinking and twisting of the major blood vessels leaving and returning to the heart which severely impairs the functioning of the heart.

The Body's Response to Shock

The tissues of the body need an adequate blood flow. To provide this, an adequate head of pressure must be available in the vessels supplying the tissues. There are basically two factors which determine the blood pressure (BP).

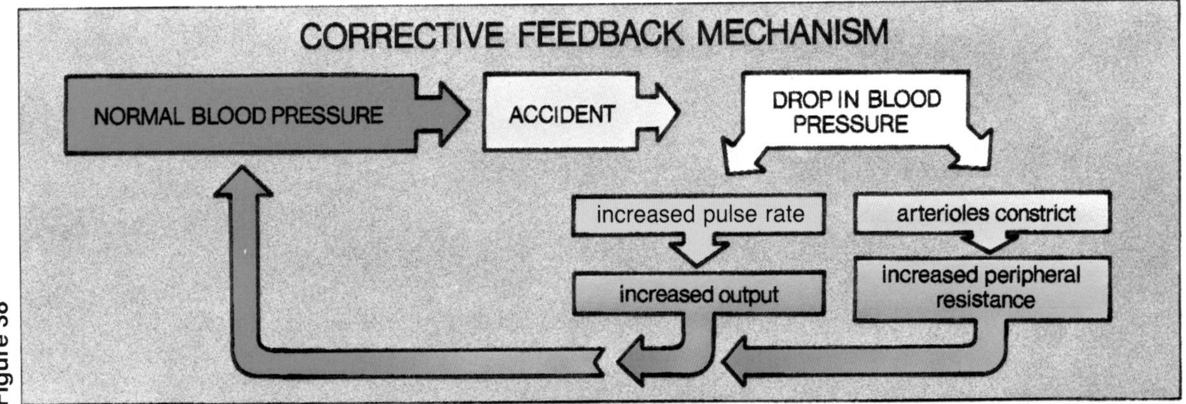

Figure 38

BP = Cardiac Output x Peripheral Resistance

The Cardiac Output is the amount of blood being pumped from the heart. The Peripheral Resistance is determined by the size of the arterioles. When they contract the peripheral resistance increases; when they relax and dilate the peripheral resistance decreases.

In cases of low blood volume (hypovolaemia) there is diminished venous return and as a result diminished cardiac output so the blood pressure then falls. In cases of low resistance shock the peripheral resistance is reduced, and so the blood pressure falls again. If the blood pressure falls below a certain critical level the tissues cease to function.

The body, however, detects that all is not well and certain compensatory mechanisms raise the blood pressure in order to sustain life. The body does this by increasing the pulse rate from the normal value of 72 beats per minute to as much as 200 beats per minute and thereby increases cardiac output. The arterioles constrict in most parts of the body, thereby greatly increasing the total peripheral resistance. The veins contract, helping to maintain adequate venous return. This is known as a corrective feedback mechanism (see Figure 38). However, if for example haemorrhage continues unchecked, a number of harmful runaway feedback mechanisms become operative which cause a vicious cycle of progressively decreasing blood pressure, circulatory collapse and death (see Figure 38).

Because of the poor oxygen delivery to the tissues the cells obtain their energy by the anaerobic process of glycolosis that leads to tremendous quantities of excess lactic acid in the blood and a consequent acidosis.

Tissues, which have a local insufficiency of blood supply (ischaemic tissues), release toxic factors; in particular the ischaemic pancreas releases a poisonous substance, which stops the muscular tissue of the heart functioning properly (myocardial toxic factor). This factor in particular, together with the excess lactic acid and degeneration products from dying tissues depresses the muscular tissue of the heart still further. In severe degrees of shock the runaway feedback mechanisms become more and more powerful, leading to such rapid deterioration that the corrective feedback systems cannot return the cardiac output to normal.

Recognition of Shock

The skin is cool and pale, and may have a greyish tinge because of a cessation of blood flow in the minute blood vessels (stasis) and a small amount of cyanosis. Profuse sweating may occur. Respiration is shallow and rapid, and intense thirst is a prominent sign. Because of a low blood pressure and inadequate oxygen reaching the brain, the person may feel faint or giddy and may appear confused. The pulse is rapid and weak. Blurring of vision may be described. Nausea is common and the person may vomit. He may appear also very anxious.

Treatment

Lay the person down and deal with the injury or underlying cause of the shock, e.g., stop further bleeding. Do not allow the casualty to sit up or stand as this further compromises the circulation. Gravity should be put to work to help rather than hinder the body's own corrective mechanisms. Raising the casualty's legs aids venous return from the lower half of the body, and improves the blood flow to the head. The head-down position causes the internal organs to press on the diaphragm, compromising adequate ventilation still further and is best avoided.

Transfer the casualty to hospital without delay. An immediate blood transfusion may make the difference between life and death.

The most important nutrient necessary to prevent deterioration of the cells and death during shock is oxygen. Administer 100 per cent oxygen if you have the equipment to do so (see Pages 104–111).

Loosen clothing at the neck or waist.

If the casualty complains of thirst, moisten lips with water but do not give anything to drink.

Stay with the casualty and carefully record his pulse and respiratory rates. In the event of a respiratory arrest, artificial ventilation will be necessary.

A casualty not responding to speech should be placed in the recovery position (see Page 49).

Do not move the casualty unnecessarily.

Alcohol should be avoided because it depresses the central nervous system.

Care should be taken to avoid overheating.

If the above first aid techniques are promptly used the first aider may prevent the development of progressive shock where the runaway feedback mechanisms take over, leading to circulatory collapse and death.

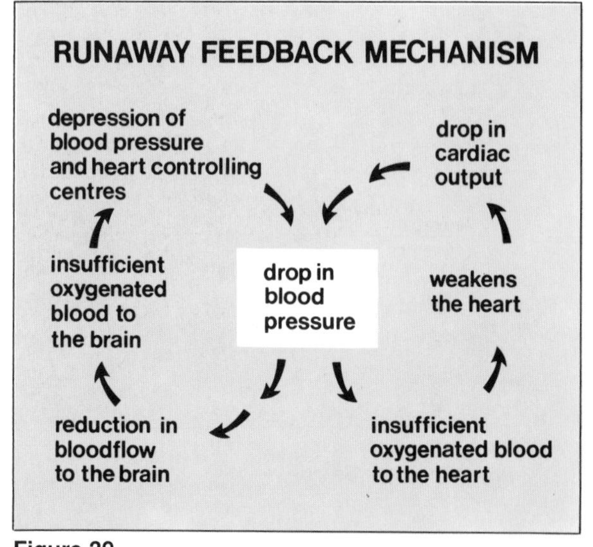

Figure 39

First-Aid Kits

First-aid kits need to be designed for the conditions under which they are likely to be used. The larger the group to be catered for and the more remote the situation, the more comprehensive the first-aid kit that is required.

The most basic first-aid kit is one which is suitable for use in a small open boat and which must suffice to return the casualty to shore for more comprehensive assistance. Because of the environment, the kit should be housed in a suitable waterproof container and should contain two large sterile dressings, two triangular bandages, a rescue blanket or large (2 x 1 metre) polythene bag and several safety pins.

Where a larger number of people are involved, e.g. the shore base area for small-boat operations or on a large diving boat, a more suitable selection of items would be:

All of these should be contained in a sturdy, weatherproof container. The preference should be for dressings of the larger sizes since a large dressing can be used on a small wound but a small dressing is of little use on a large wound. Where possible, all the items should be in individual sealed packs so those unused items remain sterile.

The most comprehensive first-aid kit in the world is of no use if no one knows how to use it. This section has scratched only the very surface of the subject of first aid and readers seeking a more thorough knowledge would be well advised to attend a recognized course of instruction such as the BSAC First Aid for Divers course or those run by the St John Ambulance, the St Andrew's Ambulance Association or the British Red Cross Society.

- first-aid instructions
- six each of small medium and large standard dressings
- large pack of assorted adhesive dressings
- three or four large triangular bandages
- ten assorted safety pins
- three 50 mm roller bandages
- 50 mm crepe bandage
- roll of 25 mm-wide zinc oxide plaster
- pair of scissors
- pair of tweezers
- pack of sterile cotton wool
- rescue blanket or large polythene bag
- disposable gloves

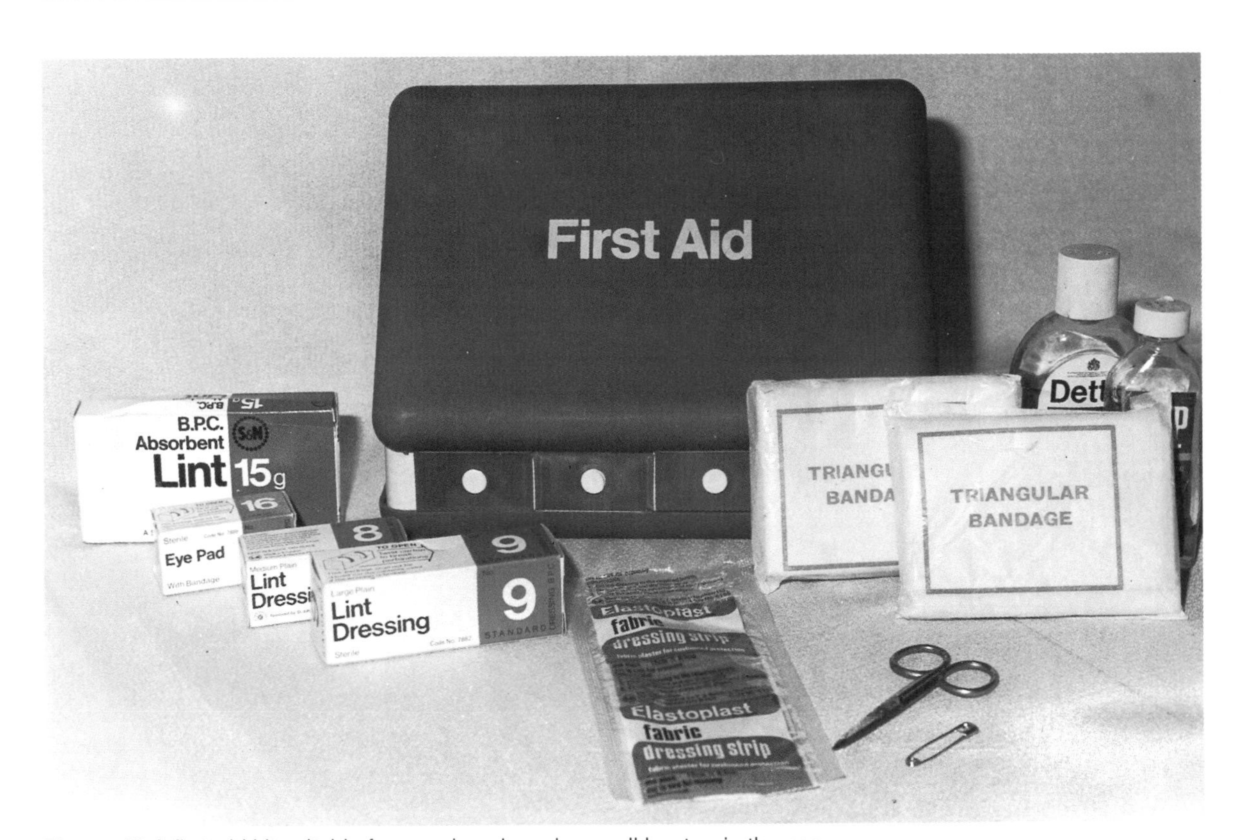

Figure 40 A first aid kit suitable for carrying aboard a small boat or in the car

Figure 41

Rescue Skills

Cardiopulmonary Resuscitation

Cardiopulmonary resuscitation comprises artificial ventilation, used to ventilate the lungs of a casualty who is not breathing, and chest compression, used to induce an artificial circulation in a casualty who has no pulse. A near-drowning casualty will not be breathing but, initially, is likely to have a pulse. If prompt action can be taken artificial ventilation may be all that is required. Should action be delayed the heart will ultimately fail and chest compressions will also be required.

Priorities of Resuscitation

Where there is **ready** access to a means of summoning the emergency services the casualty's condition should be considered. If the casualty is not breathing due to either injury or drowning carry out artificial ventilations and, if necessary chest compressions, for approximately one minute then call the emergency services.

If the casualty is not breathing due to any other cause immediately call the emergency services before commencing resuscitation.

In these circumstances the priority is to ensure that a a defibrillator, carried in many ambulances, can be brought to the casualty. The prompt availability of this equipment is thus extremely important. The subsequent chest compressions alone are unlikely to restart the heart but are vital to buy time.

In the vast majority of diving situations there will always be other people, though not necessarily divers, around. Appropriate resuscitation actions should, therefore, be commenced at the earliest opportunity while an assistant is sent without delay to call the emergency services.

In the unlikely event that diving is taking place in some remote site where there is no one else around, some compromises may have to be made depending upon the availability of a means of summoning the emergency services and the condition of the casualty.

If the site is such that reaching a means of summoning the emergency services would entail leaving the casualty for more than 5 minutes, commence appropriate resuscitation actions while trying to attract attention. As soon as assistance is obtained have the emergency services called while continuing resuscitation.

Artificial Ventilation

Artificial ventilation is the most efficient way that does not rely upon specialized medical equipment, of ventilating the lungs of a person who has stopped breathing. The technique requires little formal training to be effective, and can be applied in a very wide variety of circumstances. These characteristics make it especially suited to the diving type of accident, although the techniques are equally valuable in any other emergency. AV relies upon the rescuer blowing into the lungs of the person who has stopped breathing via the nose or mouth, and the elastic recoil of the rib cage and its muscles to complete the exhalation. The air breathed in contains approximately 16 per cent oxygen, compared with the normal 21 per cent of atmospheric air, but this is quite sufficient to maintain life. Experience has shown that success is dependent upon two principal factors:

1. **Ensure that the airway is clear and remains so.**
2. **Commence AV at the earliest opportunity – seconds count!**

AV on Land

This must also include performing AV in situations where a firm base exists, i.e., on boats. This gives the rescuer easier access, but is not essential for the success of AV.

Initial Casualty Examination

Before starting resuscitation the casualty should be examined for any signs of breathing. The chest should be watched for breathing movements as it rises and falls and the rescuer should position himself so that his face is near to the casualty's mouth and nose so that the sensitive cheek cells can detect any response. If noises can be heard from the chest there may be an obstruction in the throat or breathing passages. The casualty may also display a bluish coloration, which indicates cyanosis. This results from a lack of oxygen in the blood, and is especially easily observed at points where there are blood capillaries near to the surface of the skin, such as the ear lobes, the lips, and the inside of the cheeks. This discoloration may not be visible in cases of poisoning by agents such as carbon monoxide gas from exhaust fumes. These symptoms indicate that the brain is being starved of oxygen, and that the circulation is not functioning efficiently.

Failure to observe one of these signs must not be taken as an indication that resuscitation is not required. Delay in starting resuscitation will reduce the effectiveness of subsequent resuscitation. Starting resuscitation at an early point in the rescue will do no harm.

Clear Airway

If the casualty is laid on his back and his mouth opened wide, debris or fluid that could block the airway can be easily removed. Loose-fitting dentures may also be removed, but well-fitting dentures can be left in place. If the casualty's head is now positioned as in Figure 42, the possibility of the tongue obstructing the airway is prevented. There is no need for further action to ensure that this has been done; observations made during AV will suggest if further corrective action is required. If there is any possibility of tight clothing or equipment causing restriction of the neck, or the movements of the chest, it should be removed or loosened. It is important to check each time the casualty has been repositioned that the neck extension has been retained after any movement (see Figure 42).

Getting a Seal
Resuscitation may be successfully applied via the casualty's mouth or via the nose, the difference between these routes being primarily for the benefit of the rescuer. If the rescuer applies his mouth to the casualty's open mouth and can seal completely, then it will be found essential to seal the nose by using either the thumb and forefinger to pinch the nose, or by sealing the casualty's nose using the rescuer's cheek. Failure to seal the nose will result in air not being blown into the lungs. If the rescuer applies his mouth to the casualty's nose, the seal will be easily obtained in most cases and the mouth closed by pushing the casualty's lips together (see Figure 42). If resuscitation is required on a young baby or small child the seal may need to cover the mouth and nose for an effective seal to be obtained.

Starting Resuscitation
Resuscitation should start as quickly as possible as any delay may lessen the chances of success as explained earlier. The rescuer should blow into the casualty via the appropriate seal until resistance is met. Watching the casualty's chest will allow the rescuer to determine the efficiency of the airway that he has established. If the chest is seen to rise, the airway is open and resuscitation should continue (see Figure 43). The rescuer should then remove his mouth from the vicinity of the casualty's mouth and nose and exhale any excess air. He should then breathe in normally, clear of the area of the casualty's mouth and nose to prevent rebreathing exhaled

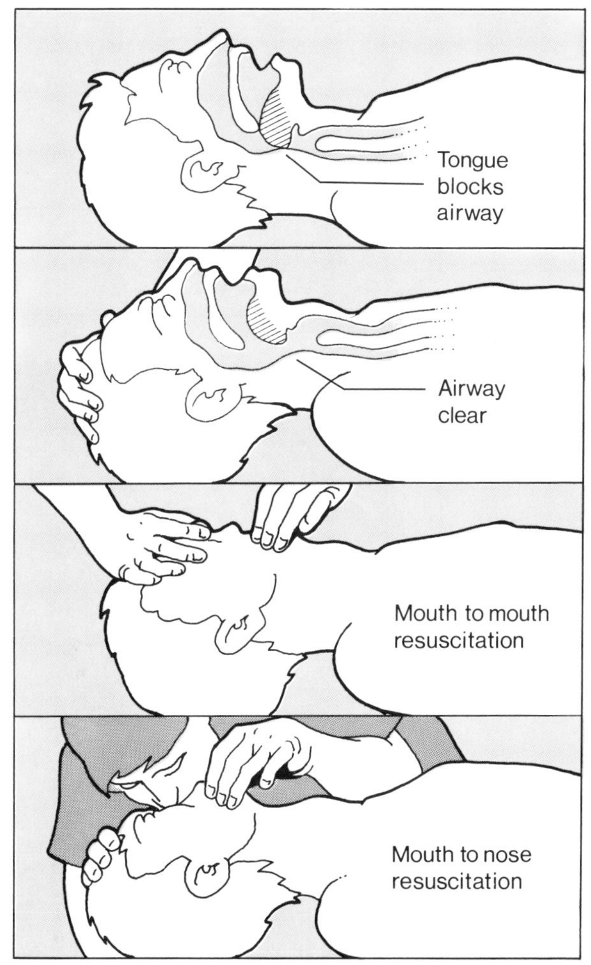

Figure 42 Extending the casualty's neck is essential if the airway is to be kept clear. Be sure to seal the opening not being used

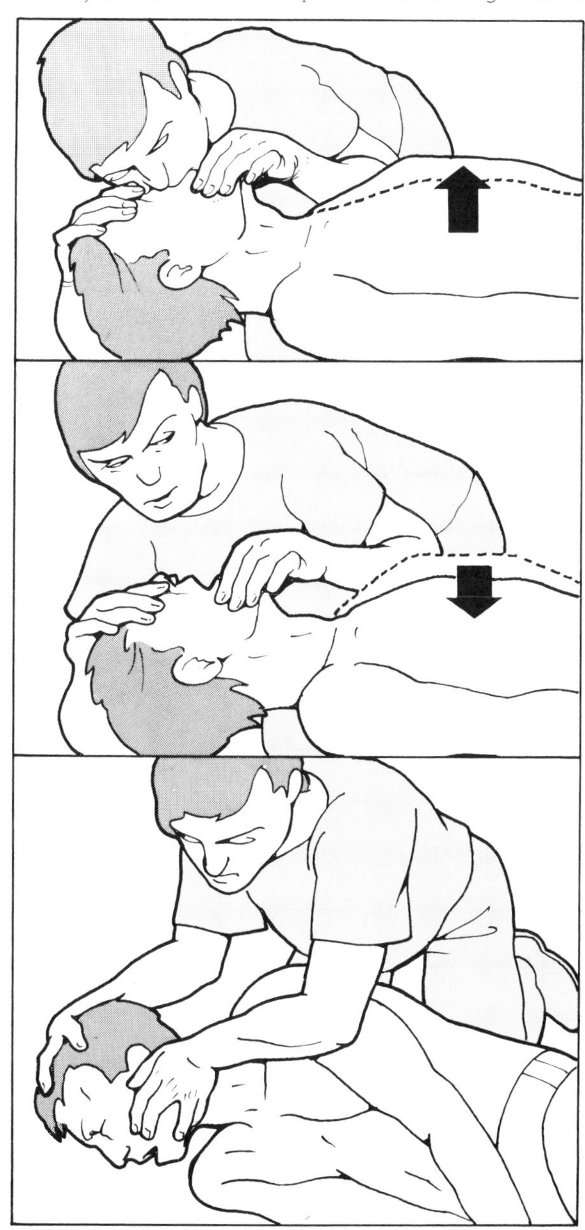

Figure 43 Check that the casualty's chest rises as an indication of effective ventilation. Should the casualty vomit, turn him over and clear the mouth before recommencing artificial ventilation

air. During this time the casualty's chest will fall due to the natural elasticity of the chest, causing the casualty to exhale passively. Once exhalation has finished the rescuer repeats the resuscitation.

Timing for Resuscitation
The effectiveness of ventilations should be judged by observing the rise and fall of the chest. The rate of ventilations should be judged by watching the rise of the chest during inflation, and then allowing the fall of the chest to dictate when the next cycle should commence.

Continuing Ventilation
Resuscitation should be continued until qualified medical advice is obtained. This may require a change of operators. Such a change can be safely performed if the new operator observes closely the procedure and then takes over at an agreed point. It is important to attempt to continue the rhythm of the ventilation. Observations made during the performance of the ventilation may require further actions, which will be discussed later.

Advantages of AV
One of the major advantages of artificial ventilation over any of the other forms of resuscitation is that it can be carried out at an earlier stage in any rescue. Due to its simple requirements it may be performed in water whether walking, swimming or supported. It can also be performed throughout a rescue using more than one operator, if others are available, with minimal inconvenience. The technique is inherently simple to learn, and has the advantage of not requiring great physical strength from the operator.

These characteristics have made it the ideal technique for virtually all situations requiring ventilation. The technique may not be appropriate, however, if very severe facial injuries have been sustained, or in cases of poisoning by corrosive chemicals, when the possible seal that could be obtained would be non-effective, or when there is a risk to the operator of contamination or injury.

Artificial ventilation in the water
Without a firm base to work from as on land, artificial resuscitation in the water is more strenuous and hence some compromise has to be achieved. Wherever possible boat assistance should be summoned, thus eliminating the need for the rescuer to tow the casualty. Even so, a lower rate of ventilations (perhaps 8 or so a minute) may have to be accepted if the rescuer is not to become exhausted. Remember that the effectiveness is far more important than the rate.

Position of the Rescuer
Prior to starting resuscitation it is important for the rescuer to inflate the casualty's BC or buoyancy aid. This should be inflated to the maximum extent that does not impair the ability to get an adequate neck extension. The casualty's mask and mouthpiece should be removed and the rescuer may find it necessary to remove his own mask if he cannot effect an efficient seal. If this is necessary the mask should be retained in case of further need during the rescue.

The rescuer should grasp the casualty in a position that will allow him to maintain a good neck extension and hence a clear airway for ventilation; the technique illustrated ensures that these requirements are met. For the sake of convenience the technique has been described and illustrated from one side only. However, it must be appreciated that either side is acceptable and that some variation is also permissible, the main requirement being the demonstrable success of the technique as regards seal and neck extension (see figures 44 and 45).

The rescuer holds the casualty by a suitable hold with one hand on the casualty's chin. This hold must remain

Figure 44 Hold the casualty's chin and place the forearm against his shoulder and use leverage to extend his airway

Figure 45 Make a good seal over the casualty's mouth for effective ventilation. This is made easier by rolling the casualty towards you

clear of the casualty's throat at all times and also supplies the seal of the casualtys mouth by gentle pressure in an upwards direction. The elbow should be kept close to the neck, passing down by the shoulder blade so that the rescuer's forearm lies close to the casualty's neck. It is important to keep the elbow below the shoulder, if it is allowed to move onto the casualty's shoulder the neck extension is greatly reduced, no leverage being possible in this position. The rescuer's other hand should be placed at some convenient position under the casualty's far shoulder blade. By pushing upward with this hand, the rescuer can cause the casualty to roll towards him thus bringing the casualty's nose close to his mouth. The rescuer can then effect a seal over the casualty's nose to administer AV with minimum need to raise himself out of the water to 'climb' over the casualty as shown in Figure 45. The less the rescuer has to raise himself out of the water the less effort he will have to expend and consequently the AV can be sustained for longer and more effectively. Rolling the casualty towards the rescuer also results in less tendency for the rescuer to push downwards on the casualty thus causing the casualty to submerge.

While the precise positioning of the hand under the shoulder blade is not critical, care should be taken to ensure that it is this hand, and not that providing the arm lever, that causes the casualty to roll. Failure to do so can result in the casualty's neck being twisted as the rescuer pulls on the casualty's chin, restricting the casualty's airway and reducing the effectiveness of the resuscitation. The arm lever should be used to ensure that adequate neck extension is maintained throughout with the other arm providing the rolling power.

AV Timing Whilst Towing

Due to the physical strain of performing resuscitation whilst towing it is not feasible to perform resuscitation at the same rate as while stationary. The procedure is therefore initially to inflate the casualty's lungs twice, and then to continue to give AV as frequently as possible, bearing in mind the need to complete the rescue. As a guide, giving two ventilations every fifteen seconds should ensure a reasonable rate of success. It is most important that the AV is effective. Failure to get the neck extension or to perform an effective seal will reduce the efficiency considerably. The main priority must be to maintain life in the casualty. Such activities as summoning assistance, removing gear or towing to a suitable landing spot should be performed so that they are integrated with the ventilations.

Resuscitation Whilst Holding On

AV may be continued successfully while holding on tc the side of a boat or a jetty or similar edge with the casualty supported by the crook of the elbow of the rescuer's right arm. AV can then be continued at the normal rate (see Figure 46).

Resuscitation Whilst Walking

AV should be continued while walking the casualty ashore (see Figure 47). The casualty can be easily supported using one hand, while the other hand is used to maintain the airway and to seal the mouth during mouth-to-nose resuscitation. AV should be continued at the normal rate as judged by watching the rise and fall of the casualty's chest throughout the landing.

Figure 46 AV while supported

Figure 47 Continue AV while coming ashore

Cardiac Arrest

Such a misfortune may be due to a variety of circumstances, including electric shock, and heart attack where the heart is affected directly, and will follow from indirect causes such as respiratory failure following drowning or asphyxiation. Unless action is taken promptly to restore the circulation, damage will rapidly follow to the sensitive tissues of the body such as the brain.

Symptoms of Cardiac Arrest

Before starting Chest Compression it is important to check that there is no circulation present. The casualty will normally exhibit the symptoms of respiratory failure already detailed, including cyanosis of the skin and lips, so the most reliable symptom is the lack of a detectable pulse. The pulse may be most easily checked at the neck (carotid pulse), in the hollow between the voice box and the adjoining muscle, a couple of finger widths below the jawbone (see Figure 48). Care must be taken to ensure that the pulse is not present since application of chest compressions to a casualty whose heart is beating may interfere with the heartbeat and cause it to become irregular or stop. Finding a pulse at other parts of the body may be more difficult, particularly if the casualty is cold. Care must be taken to ensure that a finger and not a thumb is used to check for the pulse. It is also possible to hear the heart beating if the rescuer applies his ear to the casualty's chest, however care must be taken to minimize the time taken on these diagnostic tests. Similarly, since cardiac arrest will normally accompany respiratory failure, the application of AV alone will not have caused an improvement in the casualty's cyanosis.

Performing Chest Compression

Pressure exerted over the area of the lungs and heart force blood to be circulated around the system. This is effected by pressure on the lower third of the breastbone (sternum). The relaxation of pressure allows the chest and heart to re-expand naturally and refill with blood.

To be effective the pressure must be applied to the correct place and in the correct manner (see figures 49–52). The casualty should be laid on his back on a firm surface, and the rescuer should kneel alongside and pinpoint the correct position for the application of chest compressions.

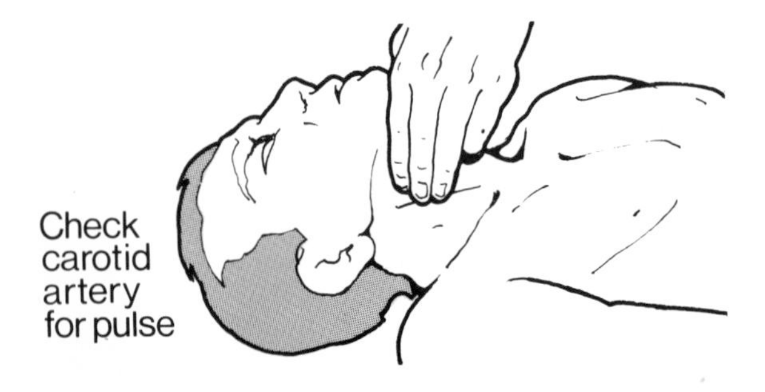

Figure 48

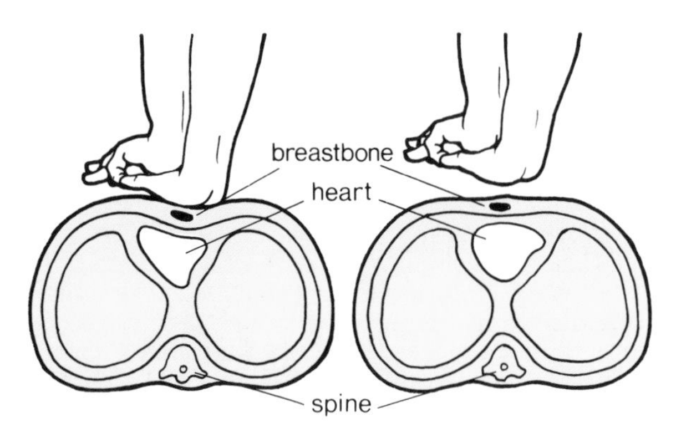

Figure 49

This is most simply achieved by locating the lower end of the breastbone and placing (see Figure 51) the heel of the palm two finger widths above this point. The other hand is then placed on top, keeping the fingers clear of the ribs.

The breastbone should be depressed a distance of 4–5 centimetres towards the spine for the average adult. For slightly built people the displacement of the breastbone should be proportionately less, 2.5–3 centimetres, and for children 1.5–2 centimetres. The heart and chest will be squeezed and will expel the blood contained within. When performing chest compressions on a child two fingers may well be sufficient to ensure the required movement. The pressure should then be released and the chest allowed to expand again. Do not interrupt chest compressions to check for a pulse returning. Should this occur it would be quite apparent because of a significant improvement in the casualty's appearance. Once a pulse has returned, check that it is still present every minute or so. If a pulse is detected the chest compressions should cease forthwith. This pressure-relax routine should be repeated at a rate of about 100 times per minute.

Cardiopulmonary Resuscitation (CPR)

Procedures for CPR depend upon whether one or two rescuers are available.

CPR for One Operator

A routine needs to be established that will allow the operator to achieve the maximum benefit for the casualty whilst not exhausting the operator. This is best done if the rate is as follows: after the initial two inflations, a sequence of fifteen compressions is given followed by two breaths of artificial ventilation. This 15:2 ratio is continued from this point.

CPR for Two Operators

If two operators are available it is useful to select the more experienced person to perform the cardiac compression. This person controls the timing of the operation and can oversee the performance. The rate that is used differs from single-operator CPR in that five compressions are given followed by one ventilation. This routine needs to be

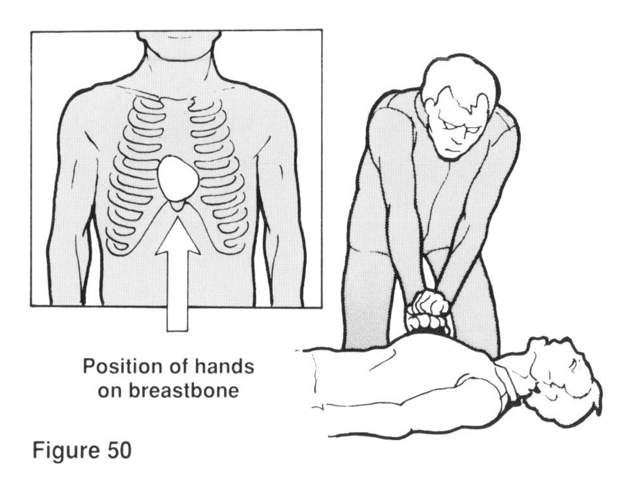

Figure 50

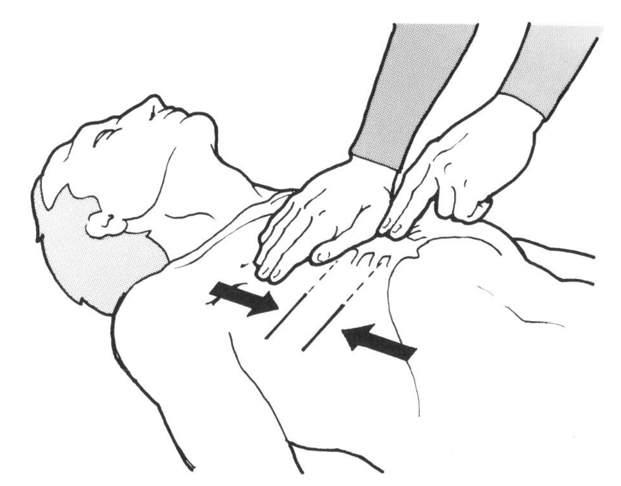

Figure 51 Locating chest compressions position

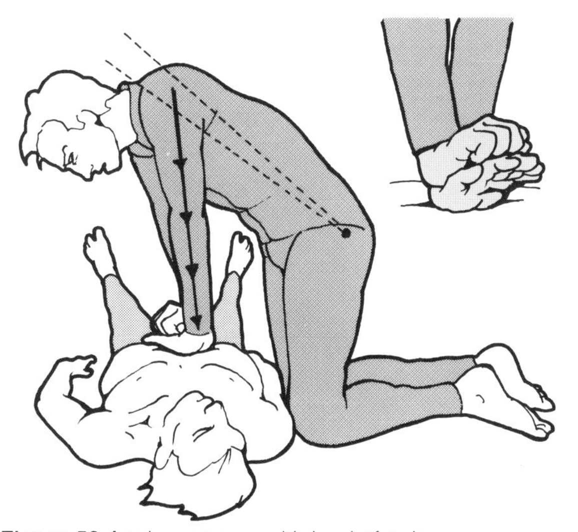

Figure 52 Apply pressure with heel of palm

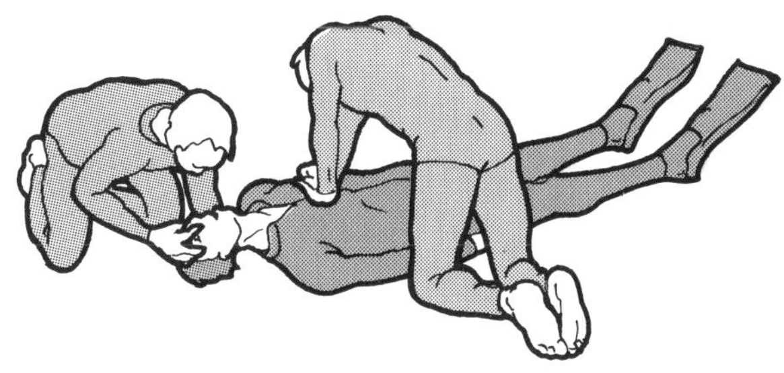

Figure 53 CPR with two operators

managed in such a manner as to produce an almost constant level of activity. The 'compressor' calls out the count from one to five, and as he completes the fifth compression so the 'ventilator' administers the one ventilation. This effectively 'blows off' the hands of the 'compressor' and ensures that the minimum time is wasted in the performance, the 'compressor' continues with the chest compressions as soon as he sees that the casualty's lungs have been fully inflated. Since this operation requires strenuous physical activity practice should ensure that it is possible to swap roles without interrupting the resuscitation. This is done under the control of the rescuer administering chest compression.

Note

Techniques for CPR **must not** be performed on a healthy person. A dummy must be used for practice. Additionally a dummy will indicate the effectiveness of the resuscitation.

The force required for chest compressions is considerable – about half the bodyweight of the casualty – and may cause rib fractures and other internal injuries. However these are less important than maintaining an adequate blood circulation in the casualty.

Figure 54

Rescue Techniques

Towing

Is a tow necessary?

For a rescue to be successful, the rescuer must complete it and be able to assist the casualty from the source of danger. The rescue may involve a variety of different activities such as rescue from depth, AV whilst swimming, landing, fetching help, etc., thus towing may form a part of the necessary process. However, it must be realized that to tow a diver in open water is a very strenuous activity and may result in the rescuer becoming incapable of further assistance. Experience has shown that the tow is the part of the rescue that creates the most difficulties in regard to personal fitness. It is therefore very prudent to seek to deal with a casualty in an alternative manner than towing for prolonged distances, and this may involve a cover boat being summoned. This will prove quicker in many cases, and will not cause the same physical exertion for the rescuer. He is therefore in a better position to be able to effect a successful rescue knowing that assistance has been summoned.

Figure 55 A conscious casualty can be towed by gripping a convenient part of his equipment

Figure 56 The unconscious casualty needs an adequate neck extension during the tow. This will require two hands

Buoyancy

If a tow is essential, some consideration must be given to the problem of buoyancy. Towing a diver is hard work, towing a negatively buoyant diver may prove impossible. The first priority must be for positive buoyancy to be generated by the rescuer. This can be done in a variety of ways and will depend upon the equipment being worn by the casualty. A BC will help ensure the correct attitude in the water but care must be taken if other actions such as AV are needed. Full inflation restricts the movement of the casualty's head with some types, and may result in a reduction in the possible neck extension. If AV is needed as much air as possible should be put into the BC, without interfering with achieving an adequate neck extension. If drysuits are being worn, these should not be relied upon to provide buoyancy at the surface as obtaining adequate neck extension frequently causes air to leak from the neck seal resulting in lost buoyancy at a critical time. Where air does collect around the neck seal this may cause some distress to the casualty by interfering with his ability to breathe. During the tow the rescuer's BC should only be inflated to the minimum necessary since this will restrict his movement and may prevent him from carrying out a full rescue effectively.

Ditching Equipment

Unless you are faced with a long surface tow it is generally not greatly beneficial to remove equipment; the main priority must be the life of the casualty. It is better to concentrate on AV and summoning assistance than ditching equipment. If, however, a long tow is necessary equipment should be ditched, since the extra drag will reduce the efficiency of the tow and hence will involve the rescuer in greater effort. Consideration should be given to the method of ditching equipment so that the minimum time is lost and interference with AV is reduced. This will normally involve ditching the casualty's weightbelt, followed by a slackening of shoulder straps, then releasing the waist strap.

Technique

Since towing requires strenuous physical effort, every consideration should be given to reducing this by correct positioning of the rescuer to casualty. The technique to be used will depend upon the condition of the casualty and the contact required by the rescuer.

Conscious Co-operative Casualty
A conscious and co-operative casualty should be towed, as should all other casualtys, from behind using a convenient grip on the casualty's BC or equipment (see Figure 55). A secure hold is desirable, but it is also essential for the rescuer to be able to release his grip in the event of sudden panic by the casualty. This position will also allow the rescuer to look where he is going regularly throughout the tow.

Unconscious but Breathing Casualty
An unconscious casualty must be protected from the risk of further drowning during the tow by the actions of the rescuer. This will involve the rescuer using two hands, one on the chin to provide adequate neck extension and one to provide a secure hold. This position will allow good neck extension to be maintained, and will allow the rescuer to assess the casualty's condition during the tow. If it becomes necessary to administer AV, the rescuer requires minimal change in position (see Figure 56).

Unconscious Non-Breathing Casualty
As has already been stated in the section on AV, a compromise is essential for the successful application of AV whilst towing. The position to be adopted is based upon the need for neck extension and the ability of the rescuer to perform a mouth-to-nose seal for AV (see Figure 45). The compromise must normally come in the towing section; if AV is administered then progress towards a distant point must be reduced. As AV is administered in a virtually static position, only minimum progress is made in order to keep the casualty's feet from dragging and entangling the rescuer's. In all of these positions, the ease of towing is considerably enhanced if a 'line-astern' position is adopted so that the rescuer-casualty group presents the most streamlined surface possible.

Pace
The rescuer must be able to maintain the pace set throughout the rescue. It is very important that the rescuer does not overtax himself early on in the tow and then find that he cannot complete it. This requires careful self-discipline to avoid 'rushing for the finishing line' and must also reflect careful evaluation by the rescuer as to his ability to tackle a rescue. At a time of stress it is all too easy for precipitate action to lead to greater problems.

Figure 57 The rescuer should fin to either side or below the casualty for greatest effect

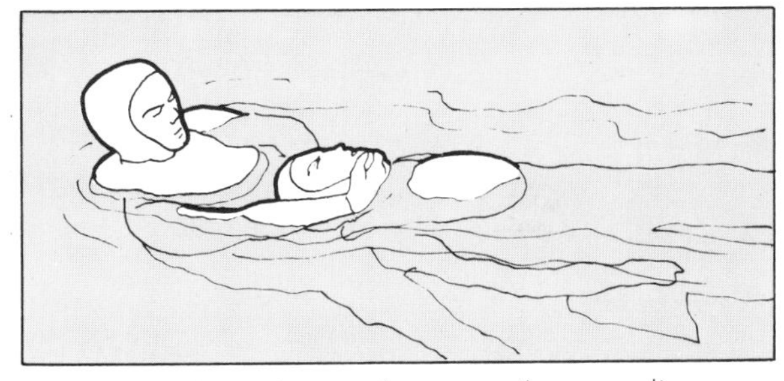

Figure 58a Conscious and co-operative casualty

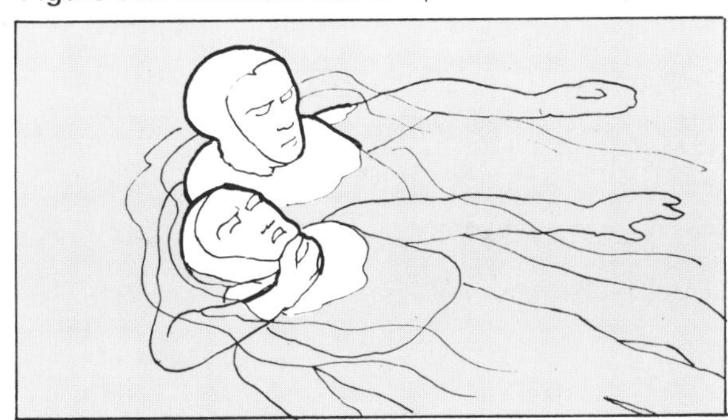

Figure 58b Unco-operative – free hand for swimming

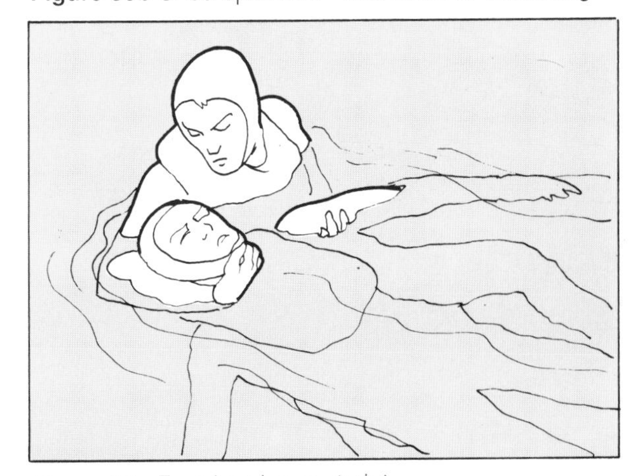

Figure 58c Free hand as restraint

Removing the casualty's BC

Figure 59 Removing BC from shoulders

Figure 60 Pulling BC clear

Aftercare

Once the casualty has been removed from the water it is important for treatment to continue until he can be handed over to medical care. This aftercare must commence as soon as the casualty has been removed from danger so that the long-term effects of the accident can be minimized. This care has already been explained in detail in the earlier section on shock (see Pages 33–34). It is useful, however, to think in terms of the management of a casualty following a diving or drowning accident and to investigate the possible points that should be followed (see also Pages 94–95).

Removing from Danger

The techniques of landing are explained on pages 72–79 and should allow the casualty to be removed from the water to a point of safety. It is important to place the casualty in a position that will not allow him to accidentally slip back into the water. The most suitable position to prevent the casualty rolling, whilst at the same time assisting his recovery, is the recovery position.

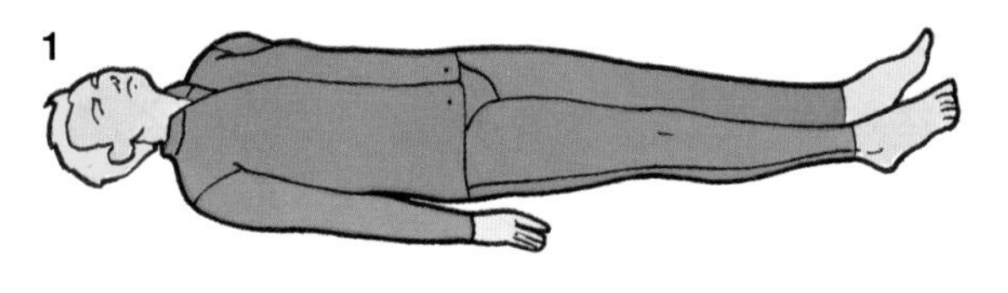

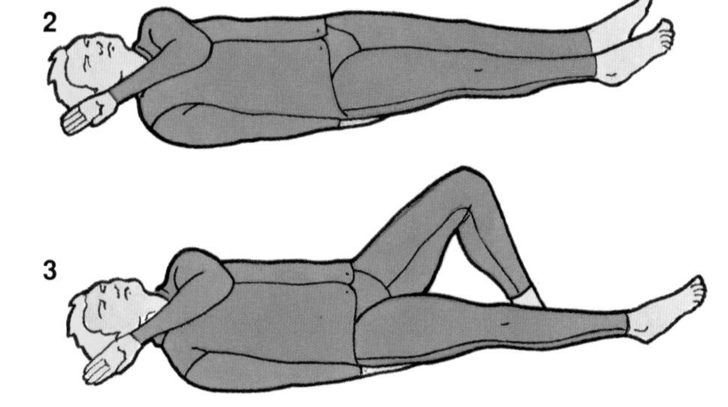

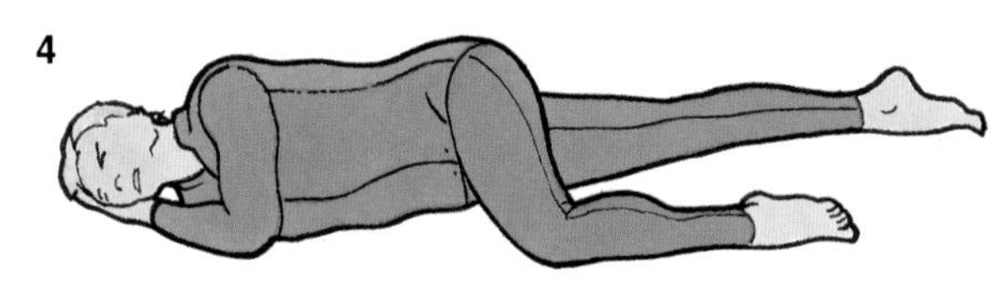

Figure 61 Sequence for placing casualty in the recovery position

Recovery Position

This position is intended to reduce to a minimum the obstructions to breathing and the risk of inhaling vomit. The casualty will lie with the chest actually not quite resting on the ground. The technique used to place someone into this position from a position lying on his back is as follows (see also Figure 61).

1. Lie the casualty's arm which is nearest to you straight along his side with the hand, palm upwards, tucked under his thigh.
2. Pull up the casualty's far leg until the foot is close to the other knee.
3. Place the casualty's far arm across his chest with the hand, palm outwards against his near cheek.
4. Using the casualty's far knee as a lever and holding the casualty's hand against his cheek, roll the casualty towards you until he is lying on his side supported by his elbow.
5. Bend the casualty's upper knee until the upper leg is at right angles to his body.
6. Ensure that the casualty's head is placed to provide a clear airway and is angled such that any vomit will naturally drain.

(a) his head put back into the extended airway position
(b) his face angled towards the ground. This will allow his mouth to be opened and allow the exit of vomit from his mouth.

Once in this position, the casualty can be protected from further heat loss using suitable clothing, plastic bags or exposure blankets and should be transported in this position if possible. Careful observation should be maintained to ensure that the pattern of breathing is not interrupted. If required the casualty can be quickly rolled back into a suitable position for cardiopulmonary resuscitation (CPR) to be performed.

Treatment of Injuries

Priority should be given to breathing and circulation, and other injuries should only be tackled once these have been stabilized. If, however, the injuries are very severe and threaten life then appropriate action must be taken to reduce the risk of their worsening. Severe bleeding may require the use of direct pressure (see Pages 30–31). Other injuries may be tackled only if treatment does not reduce the efficiency of CPR if it is being applied.

Use of Oxygen

If oxygen administration equipment and expertise is available then oxygen should be freely given if there are no contra-indications. It is more beneficial to use oxygen heavily for a short time than lightly for a long time. For further information on the use of oxygen (see Pages 104–111).

Swimming Rescue

There are circumstances where the rescuer is the only person available to respond to an emergency situation. In this situation it is even more important to ensure the safety of the rescuer, no matter how skilful or fit he may be. The following methods of surface rescue should be considered in this order:

> **REACH**
> **THROW**
> **WADE**
> **ROW**
> **SWIM (with a rescue aid) and TOW**

Reach
This method of rescue assumes that the casualty is close to the shore or boat. The use of a rigid aid can increase the reach of the rescuer, e.g., paddle, branch or stick. Clothing can also increase the effectiveness of this method (see Figure 62).

Throw
Throwing a buoyant aid is a first step in a rescue attempt, and generally only provides the person in difficulties with additional buoyancy whilst other assistance is being summoned.

Throwing a rope can also be an effective method of rescue. The success of this method will depend on the rescuer's accuracy and to some extent on the type of rope being used.

Wade
Wading is a much quicker method of approaching someone if the conditions are suitable. A sure foothold will assist the safety of the rescuer, and objects may be taken by the rescuer to assist him in stretching out towards the casualty, thereby extending his reach.

Row
If reach and throw rescues are not possible due to the distance to the casualty, and a wade rescue is not possible because of the depth of water, a rowing boat or any other type of surface craft may have to be used. However, care must be exercised to ensure that the boat is suitable and safe for the rescue, that the rescuer is familiar with the type of craft, and has the necessary skills required. If a small craft is being used, there may be difficulties in landing someone into it without capsizing the boat; this may require bringing them in over the transom.

Swim and Tow
The principle is to reduce the degree of risk taken by the rescuer. Swimming rescues should only be undertaken if all other methods are inappropriate. Tows are covered on pages 46–47.

Reach

Throw

Wade

Row

Swim

Tow

Figure 62

Divers are more likely to need to perform their rescue skills on a fellow diver in distress. However, divers are not the only users of the water environment; swimmers, canoeists, sailing enthusiasts and any other water users may also get into difficulties. Additionally, non-water users who did not intend to go into the water can get into difficulties. This group accounts for the majority of drownings, accidentally falling into rivers, lakes and ponds, etc.

Entries

Slide-in Entry
The slide-in entry should be used where the depth of water and type of bottom are unknown. This method allows the feet to feel for any unseen obstacles. The rescuers should lower themselves into the water, taking the weight on the hands. A forward-facing entry allows the rescuer to watch the casualty throughout the entry (see Figure 63).

Step-in Entry
This entry is used when the depth of water and type of bottom are known. The rescuer should step out (not jump), keeping the knees slightly flexed in order to give way when the bottom is detected.

Stride Entry
The stride entry is used when the rescuer needs to keep the casualty in view and when the ground is not much above the water level. Unlike the slide-in method, it is assumed that the depth of water and bottom conditions are known. The rescuer should step out with one leg extended and the other leg backwards, leaning forward with the arms extended and palms down. On entering the water press downward with the arms while keeping the head erect.

Jump Entry
This entry is only made when the depth of water is known, for example from a charter boat where the height is too great to slide in. The disadvantage with this method is the loss of visual contact with the casualty.

Dive Entry
As with the jump entry the depth of water must be known. However, the shallower the dive entry the greater the distance covered by the rescuer.

Figure 63

JUMP ENTRY
STEP-IN ENTRY
STRIDE ENTRY
DIVE ENTRY

Approaching the Casualty

In an emergency speed is essential, although the rescuer must pace himself in order to reach the casualty in a sufficiently fit state to effect the rescue. If the depth does not allow a wading approach then the most effective swimming method is the breaststroke. The breaststroke, although slower than other methods, does allow for good forward vision, especially if the rescuer is pushing a buoyant aid. Most divers will have a diving mask and fins available which will enable them to reach the casualty more quickly, albeit in a head-down position. However, it is important that as the distance to the casualty reduces, a head-up approach is adopted so that encouragement and assurance can be given (see Figure 64).

Figure 64 Approaching the casualty

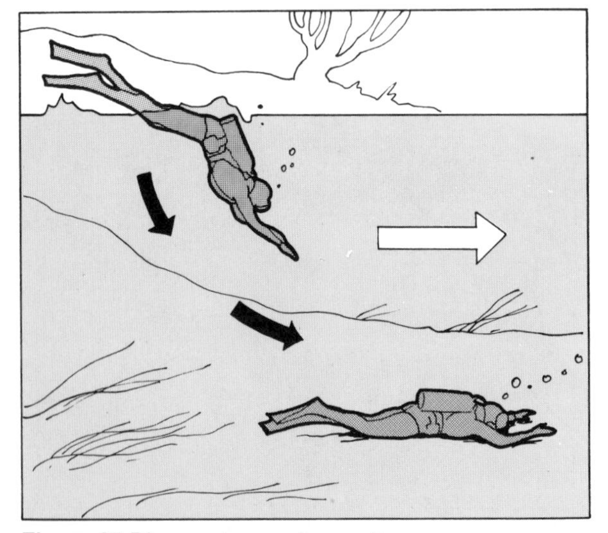

Figure 65 Dive upstream of casualty

Figure 66 Release from neck-hold

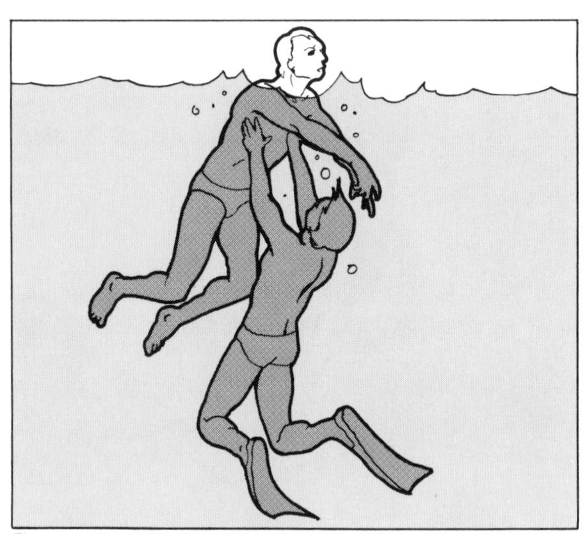

Figure 67

Rescues in rivers or water where there is a strong current or flow should be planned so that the entry is at a point where the current can be used to provide the shortest distance to the casualty. The rescuer should avoid making immediate contact with the casualty, staying just out of reach until he has had time to assess the situation. If the casualty is passive then the rescuer can inflate their BC or give them physical support before towing (see Pages 46–47).

If the casualty is struggling and panicking then the rescuer should try to calm him down. Offering a buoyant aid, and allowing the casualty to grab it, is a non-contact method of rescue. If the casualty is unconscious then contact must be made immediately. Speed is essential.

If the casualty sinks during the approach the rescuer will have to perform a surface dive in order to locate the casualty. This can be achieved either with a head-first or feet-first dive (see Figure 65). Remember, if the water is moving you will have to allow for drift. Bubbles on the surface may indicate the position of the casualty.

Defensive Position
During the final approach the rescuer should adopt a defensive position by bringing his legs forward while still keeping out of range (see Figure 69). This method will allow the rescuer to move away from the struggling casualty while still keeping him in vision.

Blocking Methods
Should the casualty make a lunge or try to grab the rescuer, then the rescuer should push him firmly away with his leg, or raise an arm to block the attempt. With the arm block it may be necessary to submerge briefly in order to back away (see Figure 70).

Escape Methods
If the rescuer is unfortunate enough to be caught by a struggling casualty, there are a number of methods of escape.

Most grasps are around the front, usually about the neck or head. The rescuer should tuck his head down and submerge, at the same time pushing upwards and hard against the casualty's arms or chest while backing away (see figures 66 and 67).

Although being grabbed from behind is less frequent, the rescuer should again tuck his head down and locate the casualty's elbows or arms and push upwards (see Figure 68).

To escape from a wrist grasp the rescuer will need to force his arm first up and then down vigorously, while at the same time using his other hand for extra strength (see Figure 71). Again it may be necessary to submerge the casualty in order to get free. Further leverage can be obtained by placing a foot on the casualty's chest and pushing hard.

The escape method for a leg grasp is basically to stand up and push the casualty firmly away with the arms.

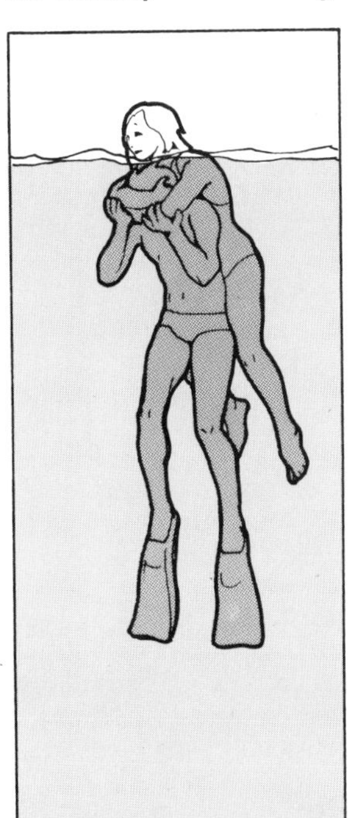
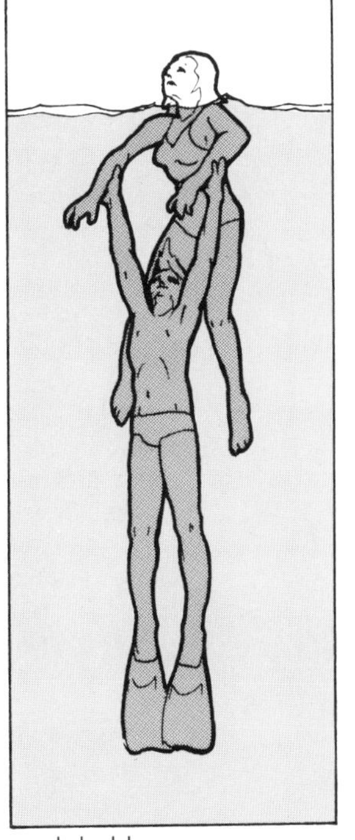

Figure 68 Release from back neck-hold

Figure 69 The defensive position

Figure 70 The defensive block

Figure 71 Release from a wrist grip

Self-Rescue

In an ideal world good diving practices would not only ensure that a diver and his partner are always in visible range of one another but also that, in good visibility they are always within assisting range of one another. In the real world it sometimes happens that, for a variety of reasons, this is not always the case. Should a separation be followed by some type of incident involving one of the divers before contact could be re-established with his partner, then that diver will have to rely on his own resources to regain the surface.

By its very nature, self-rescue can only be performed by a conscious diver and is usually as a result of a cessation of the diver's air supply. This is sometimes caused by a mechanical failure of the diver's breathing apparatus, but may also be caused by the diver running out of air due to inadequate monitoring of his air supply during the dive.

There are three self-rescue techniques that will enable the diver to regain the surface. These are the use of a secondary air supply, the controlled use of buoyancy, provided by either a BC or drysuit, and the 'free' (swimming) ascent. As the latter two techniques provide only the motive power for the ascent, a technique of breath control necessary to counteract the effects of the reduction in pressure that the diver will experience during the ascent is necessary. This is common to both and will be discussed further before the latter two ascent techniques are described.

Alternative Air Sources

Depending upon the configuration of the diver's equipment, should his air supply fail it is possible that alternative air sources are available to him

The most reliable alternative would be a totally independent breathing system. This can be provided by one of two methods. The diver may carry a small 'pony' cylinder of air with a second regulator, or if the diver is wearing a twin cylinder set, each cylinder may be equipped with a separate regulator. Failure of the primary air supply would merely require a change to the secondary supply. A normal ascent to the surface could then be performed.

A second regulator attached to the primary air supply by an adapter yoke will protect against regulator failure but not against exhaustion of the air supply. Similarly an 'octopus' second stage will only protect against failure of the regulator's primary second stage.

Many BCs are fitted with diaphragm-operated demand mouthpieces to draw air from the bag of the BC. The techniques of clearing water from the mouthpiece and hose (especially for a diver who has already exhaled) and the complication to buoyancy control require constant practice if they are to be effective. Rapid depletion of the BC's limited air supply (particularly at depth) can result. As this same air may be required to provide buoyancy for the ascent, it is recommended that a positive ascent be the first priority.

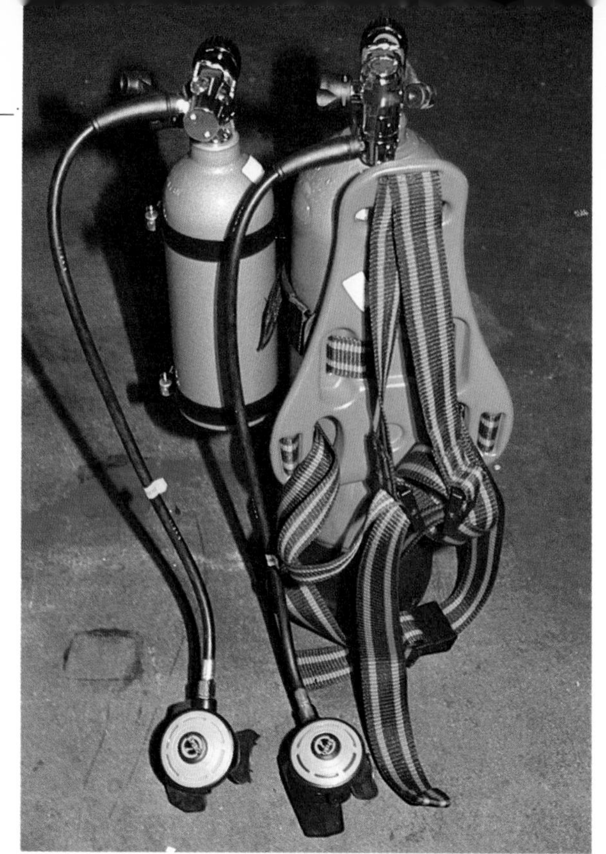

Figure 72 Cylinder with 'pony cylinder' attached

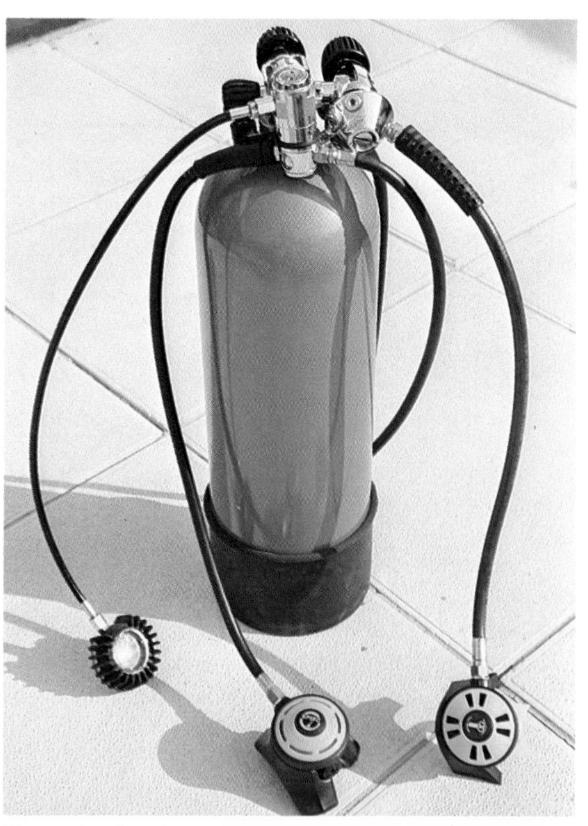

Figure 73 An alternative air source – second regulator

Breath Control

When deprived of an air supply while submerged, the diver's natural reaction will be to hold his breath. In the diving situation this is precisely the wrong thing to do and consequently, in order to regain the surface safely, the diver must control this natural reaction. The exercise of this control will only be effective if the diver fully understands what is required and why.

A failure of the air supply is almost inevitably detected when the diver attempts to breathe in following a normal exhalation. At this point a diver's lungs will still contain some air, equal in volume at least to the 'residual volume' of the lungs and usually slightly more. At the commencement of an emergency ascent, therefore, the diver's lungs will not be completely empty of air. As the ambient pressure falls during the ascent, this residual air will expand and, were he to take no action to prevent it, the diver would run a severe risk of experiencing a burst lung. To prevent this, therefore, the diver must exhale during the ascent to continually counteract the expansion of the air.

Because at the start of the ascent the diver's lungs will not be full, it may be more comfortable for the diver to hold his breath initially. The lack of a reliable reflex in the human body that will indicate to a diver that his lungs are full to capacity means that any period of breath holding must, for safety's sake, be brief. This is especially important if the ascent starts in shallow water, where the rate of expansion of the air in the diver's lungs will be greatest.

If for any reason the diver starts the ascent with lungs that are substantially full, for instance after having taken a breath of air from another diver's regulator, then exhalation must start as soon as the ascent is commenced.

The rate at which air is exhaled should initially be akin to that of a gentle whistle. As the ascent proceeds, the rate of exhalation should be progressively increased to a forceful blow as the surface is reached.

The build-up of the desire to breathe, normal during periods of breath holding, is relieved during the ascent. This is due to the fact that the build-up of carbon dioxide level in the lungs, which provides the stimulus for breathing, is prevented by the constant exhalation eliminating excess carbon dioxide from the body. The experience of divers who have had to perform these types of ascents has shown that this makes the ascent considerably more comfortable than would otherwise be expected.

Figure 74 Ascending while breathing from BC

Controlled Buoyant Ascents

This type of ascent utilizes the controlled inflation of a BC or drysuit to lift the diver to the surface at a safe steady rate. This requires that the diver be completely familiar with the controls of his equipment. The choice of which to use depends on the configuration of equipment worn by the diver and the serviceability of his air supply.

A wetsuited diver will only have the option of using his BC. A drysuited diver, however, could use either his suit or his BC. The choice in this case should preferably be that system which the diver has used to adjust his buoyancy during the dive. This will avoid the problem of having to control two volumes of air during the ascent, one in the suit and one in the BC. As most drysuits and some BCs are inflated by direct feed systems, this option may be precluded if the diver's air supply is exhausted.

To commence the ascent, the diver introduces air into his BC or drysuit until slight positive buoyancy is attained and upward motion is achieved. The diver then commences to exhale as described above. Once the rate of ascent matches that of the diver's small exhaust bubbles the inflation is stopped. Throughout the subsequent ascent excess air is vented from the BC or drysuit to maintain a constant rate of ascent. This is easier to control if the venting is done as a series of short 'blips' rather than varying the rate of a continuous bleed off. If too much air is vented the diver will start to sink again and further inflation of the BC or drysuit will quickly be required to re-establish the ascent.

Some drysuits are fitted with automatic dump valves. If set correctly any excess air introduced into the suit to increase buoyancy would be automatically vented off before upward motion began. The diver has two options to overcome this. The valve body may be screwed in half a turn to increase the blow-off pressure and hence the diver's buoyancy or the diver may opt to use his BC, leaving the dump valve to maintain a constant air volume in the suit. In practice automatic dump valves vary markedly in their characteristics and while some will cope with the volume flows required others would not, resulting in an accelerating ascent unless another means of venting air from the suit is used. The lesson is clear – know your own equipment!

Initially little venting will be required but, as the surface is approached, more frequent venting will be required to maintain a constant ascent rate. Immediately on reaching the surface the diver should fully inflate his BC to provide maximum surface support.

Figure 75 Controlled buoyant ascent, BC/wetsuit

Figure 76 Controlled buoyant ascent, using drysuit

Figure 77 Emergency regulator

Figure 78 A controlled buoyant lift

Free Ascents

A free ascent is performed when, for some reason, a controlled buoyant ascent cannot be performed. The ascent is commenced by the diver finning upwards and exhaling as described earlier.

If the diver was trimmed to neutral buoyancy before the start of the ascent, the effort required to commence the ascent will be minimal. Once the diver has ascended a short distance, the expansion of any air in the diver's BC or drysuit, or of the suit material itself, will increase the diver's buoyancy and the diver will be able to cease finning, as the ascent becomes a buoyant ascent. The remainder of the ascent will proceed as described for the controlled buoyant ascent.

If the diver was not trimmed to neutral buoyancy before commencing the ascent the initial finning may require substantial effort. This will quickly use up the diver's precious oxygen reserves and prejudice the success of the ascent. The diver should therefore jettison his weightbelt or any other heavy equipment that is easily disposable (e.g., tools or cameras) to reduce this negative buoyancy and hence reduce the effort required.

Weightbelts should not, however, be just released and allowed to drop as experience has shown that weightbelts released in this way often become snagged on other items of equipment (knives, etc.). Once unfastened the weightbelt should be pulled clear of all equipment and held at arms length before being released to ensure that it falls clear.

Jettisoning the weightbelt should ensure that at some stage in the ascent the diver will become positively buoyant. Depending upon the configuration of the diver's equipment this may happen immediately, or may be delayed until the diver has finned up a short distance. Once this happens, however, there is little that the diver can do to control his rate of ascent and he must be prepared for a faster than normal ascent. Spreading his arms and legs and angling his fins to create the most drag will all help to reduce the ascent rate. It is essential that the diver significantly increases his exhalation to counteract the more rapidly decreasing ambient pressure as the ascent accelerates.

Uncontrolled Ascents

Both of the above types of ascent require a significant degree of self-control by the diver if they are to be performed successfully. Because self-rescue is performed in a solo diver situation there will be no second chance and the attempt must result in the diver regaining the surface. If the ascent gets out of control and a very rapid and uncontrolled ascent ensues, such as would happen if a BC became fully inflated at depth, the diver will still reach the surface. Here other divers could see and assist him and any injuries he may have sustained in the ascent could be treated. If, on the other hand, the failed ascent results in the diver sinking to the bottom there can only be one outcome.

Figure 79 Jettisoning a weightbelt

Figure 80a Surfacing following an uncontrolled ascent

Figure 80b

Assisted Ascents

The techniques covered in the Self-Rescue section describe how a diver can regain the surface by his own devices if he runs out of air. With an effective buddy system, the diver will be able to rely on sharing his buddy's air supply. This will allow a more controlled and less hurried ascent with a consequent reduction in the risk of burst lung. Should the incident occur where an obstructed surface makes a direct ascent impossible (for instance when in a wreck), the benefits will be obvious. The techniques for sharing an air supply are known as assisted ascents, and depending upon the type of equipment being shared, a number of alternatives are possible. Experience has shown that techniques which allow the rescuer to remain breathing from his own regulator while offering the distressed diver the sole use of an 'Octopus' or an alternative air source are easier to perform and much less stressful. These techniques are therefore much preferred over techniques involving the sharing of a single regulator between rescuer and distressed diver.

Figure 81 *I have no more air* The signal is made by moving the arm with the hand outstretched in and out from the throat. On seeing this signal, close with the diver making it and share air from your regulator

The 'Octopus' Rig

The addition of a second second stage to a regulator, for use by another diver whose air supply has failed, is the most straightforward of the assisted ascent techniques. During normal diving the additional second stage should be secured in such a manner that it is easily accessible to another diver and is highly visible.

During an emergency, if the divers are in visual contact, the distressed diver should give the 'I am out of air' signal and both divers should close together, the assisting diver removing the additional second stage from its stowage ready for use. By the nature of the problem it will not have become apparent to the distressed diver until after he has breathed out and subsequently tried to breathe in. Consequently his lungs will not contain a lot of air for purging the water from the regulator. The regulator should therefore be offered so that the recipient has clear access to the purge button and can control the purging himself. Alternatively, the assisting diver should gently purge the regulator as he offers it. This should be done very carefully as the high purge flow rates of many modern regulators can produce sufficient quantities of bubbles to obscure the view of the regulator, and sufficient force to dislodge a face mask.

Should the assisting diver be looking in another direction at the crucial time the distressed diver should not waste time trying to attract his attention but merely approach him, take the additional second stage from its stowage and breathe from it.

Because the two divers are now constrained to be no further apart than the length of the additional second-stage hose, they should take a positive hold on each other by grasping a substantial piece of equipment, BC harness or cylinder valve to ensure that their relative positions are maintained. Once the distressed diver has regained his breath and composure a normal ascent can be commenced. During the ascent the assisting diver should closely monitor his air supply, as it will now be being used up twice as fast. The positive hold should be maintained throughout the ascent to avoid the divers drifting apart and the regulator being jerked from the distressed diver's mouth. Buoyancy control should be carried out to maintain a normal ascent rate.

Alternate Air Supplies

The techniques to be used with other forms of emergency air supply will be similar to those described above.

Shared Regulator

Where the assisting diver has only a single second stage regulator on his cylinder valve, the two divers will need to adopt the regulator sharing technique. The distressed diver should attract his partner's attention and give the 'I am out of air' signal. Both divers should close together with the distressed diver taking up a position by the shoulder opposite to that over which his partner's regulator hose is routed, i.e., if the assisting diver's regulator hose passes over his right shoulder, he should position himself such that the distressed diver is by his left shoulder. This is to ensure that the regulator can be passed easily between the two divers with the minimum of movement and that it remains the right way up. As the divers close they should take a positive hold on each other to maintain the position.

Having taken a breath from his regulator, the assisting diver should then remove it from his mouth and, retaining a positive hold on it, place it close in front of the distressed diver's face where he can clearly see it. The considerations regarding purging the regulator discussed for the octopus technique apply equally to the commencement of this technique. The distressed diver then places his hand over the assisting diver's hand and guides the regulator into his own mouth to breathe from it. The assisting diver should not try to put the regulator into the distressed diver's mouth for him. It is essential that the assisting diver retains hold of

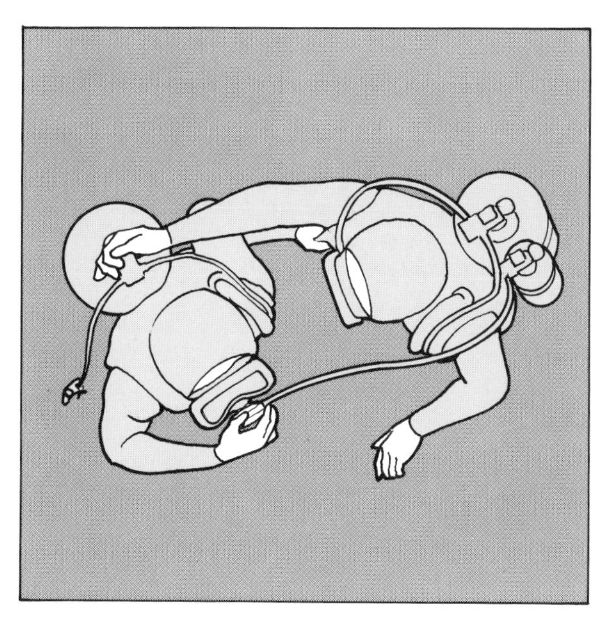

Figure 82 Pony cylinder being used for an AAS

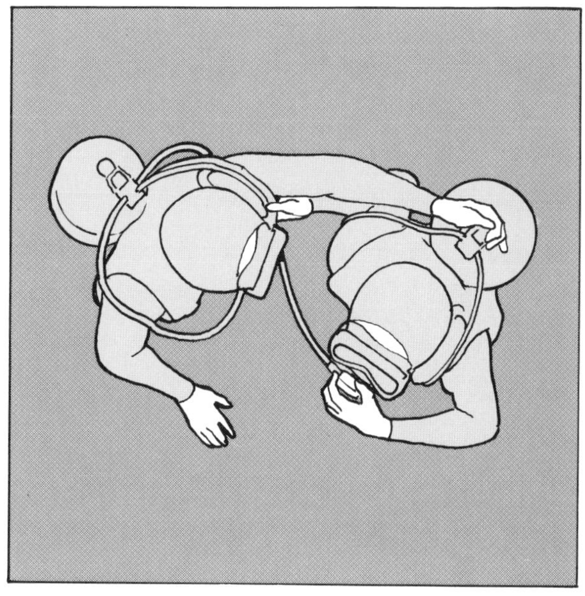

Figure 83 Using an octopus rig

the mouthpiece, and hence remains in control of the air supply throughout.

The aim is now to establish, as soon as possible, a steady rhythm where each diver takes two breaths from the regulator between exchanges. Initially the assisting diver should be prepared to allow the distressed diver more than the initial two breaths until he has regained his composure. In extreme cases the regulator may not be returned and the assisting diver will have to decide whether to forcibly retrieve it or to make a free ascent by breathing out continually during their ascent.

Once a stable rhythm has been established the assisting diver should signal 'Go up' and the pair should commence an ascent at the normal rate, maintaining their positive hold on each other and their relative positions. During the periods when each diver is not breathing from the regulator, he should gently exhale to release air expanding during the ascent and hence prevent any possibility of pressure damage. The assisting diver should check that the distressed diver is remembering to do this and should be ready to give him a reminder by means of a visual indication or by gentle pressure on his stomach.

During the ascent normal buoyancy control will need to be carried out and the assisting diver should monitor the rate of ascent, particularly as the surface is approached. Should the ascent accelerate unduly he should be prepared to overcompensate with his own buoyancy to hold back the distressed diver.

Figure 83a Divers carrying out an AAS ascent

Figure 84 Maintaining a firm grip during an AAS ascent is essential

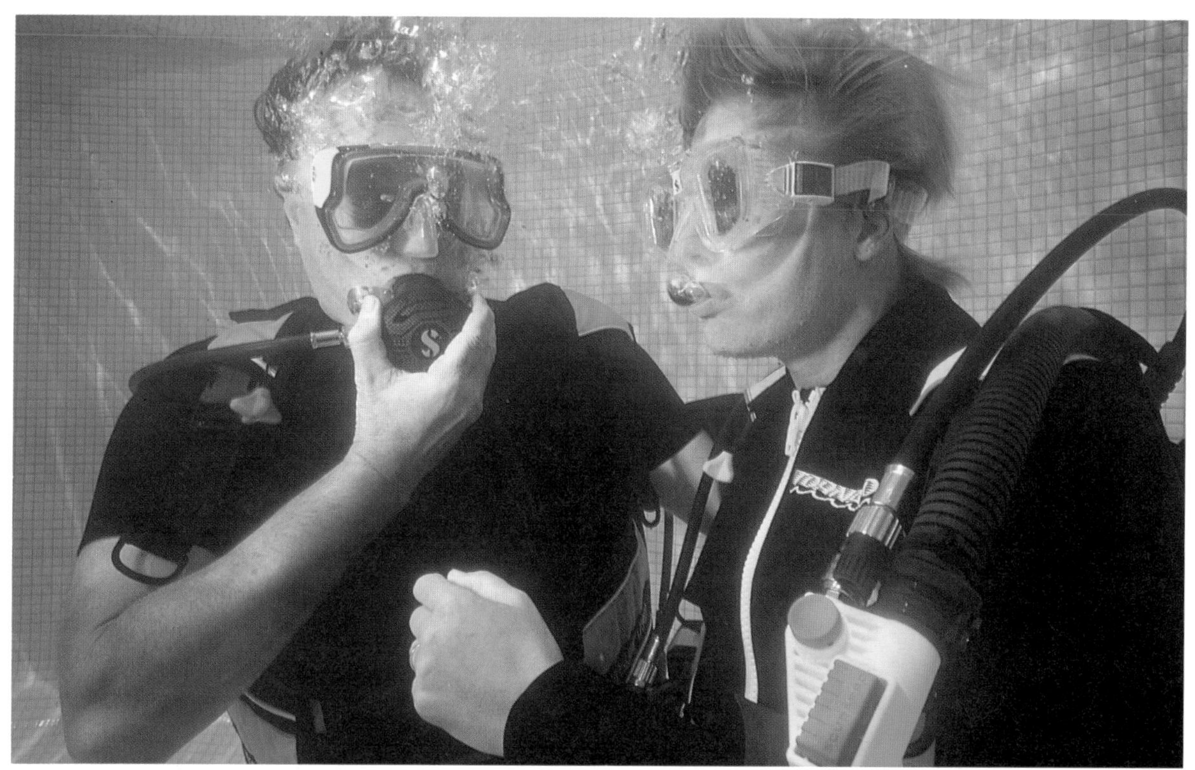

Figure 85 Diver exhales while air sharing using a single regulator

Figure 86 Donor retains grip on regulator

Rescue from Depth

In this section we consider the rescue of an incapacitated diver from depth. Such incapacitation may range from disorientation and vertigo to unconsciousness and cessation of breathing. As little can be done to treat the condition underwater, the immediate problem is to return the casualty to the surface. By the very nature of the conditions concerned this is something the casualty will not be able to achieve unaided. The rescuer's task is, therefore, to raise the casualty to the surface while not putting his own safety at risk.

The precise technique to be used to lift the casualty will depend very largely on the configurations of the casualty's and rescuer's equipment. The pre-dive equipment check is an essential action to ensure that each diver is familiar with his partner's equipment and to allow the options available for a rescue to be considered before the dive commences.

In general terms the options will fall into one of the following categories.

a) a Controlled Buoyant Lift using the casualty's buoyancy device
b) increasing the casualty's buoyancy by releasing his weightbelt
c) a Controlled Buoyant Lift using the rescuer's buoyancy device
d) a swimming lift

The rescuer should choose that method which provides the easiest means of rescue and which requires the least effort on his part. The easier the technique, the more likely it is to be successful. The less effort required the better condition the rescuer will be in to continue the other rescue actions necessary once on the surface.

The Controlled Buoyant Lift Using the Casualty's Device
This option utilizes the controlled inflation of the casualty's buoyancy device to lift the casualty to the surface in much the same way as the diver's own buoyancy device is used in the Controlled Buoyant Ascent described in the section on Self-Rescue. The considerations involved are the same, except that the rescuer must now apply them to the casualty's equipment rather than his own.

At the start of the rescue, the rescuer must establish a positive hold on the casualty to ensure that contact is maintained throughout the rescue. In most cases this will require the rescuer to be face to face with the casualty in order to have access to equipment controls. Caution should, however, be exercised when approaching a

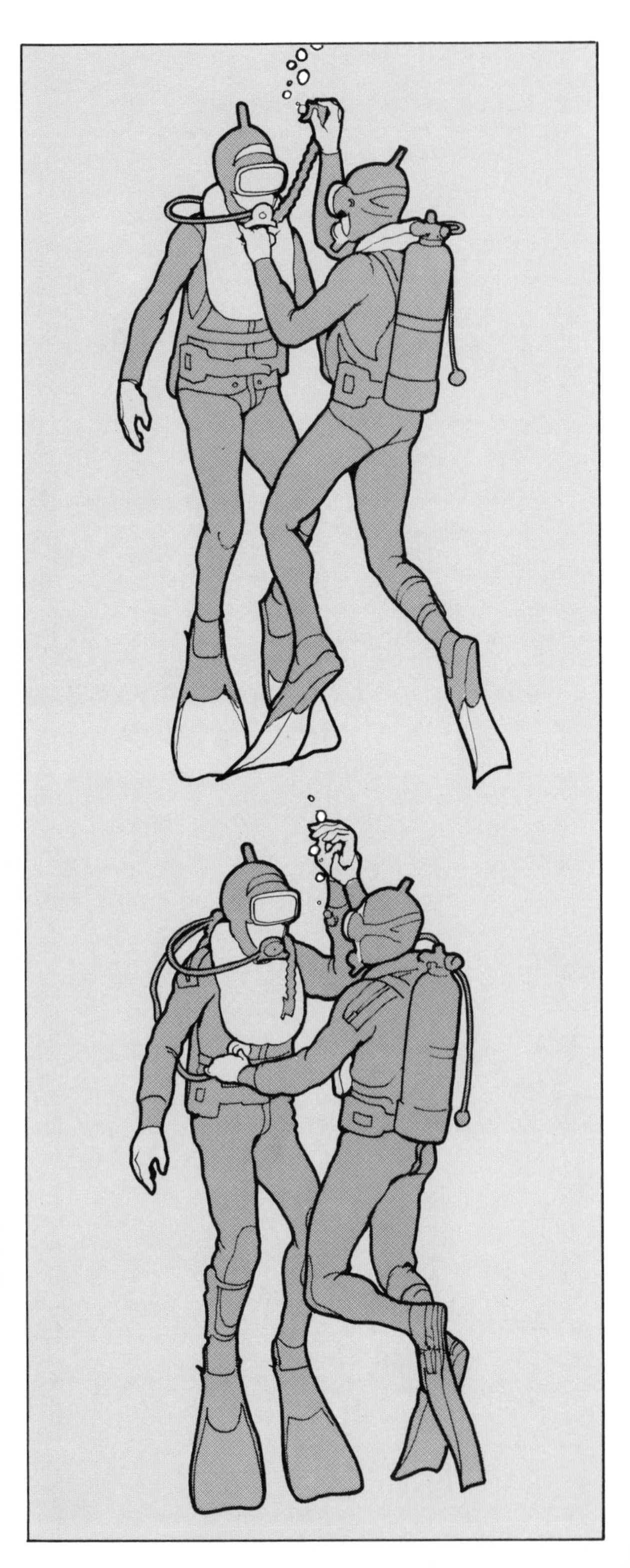

Figure 87 It the casualty is wearing a wetsuit, make positive contact and control ascent by inflating or venting the casualty's BC *(top)*.
If the casualty is wearing a drysuit, then venting may be possible or necessary via a wrist seal or wrist mounted valve. Otherwise locate and operate the dump valve. The other hand will be required for inflation via the suit inflator

casualty suffering from severe nitrogen narcosis. An approach and hold from the side should still give adequate grip and access to controls while keeping the rescuer in a position from which he can retreat should the casualty, in his narcotized state, try to take any action which may be prejudicial to the rescuer's safety.

The hold on the casualty should be taken so as to minimize the necessity to change grip during the ascent. This should also take into account the need for access to the rescuer's own equipment controls to regulate his own buoyancy.

For a casualty wearing a BC a one-handed grip around the neck of the stole or on some convenient frontal harness is effective. This leaves the second hand free for inflation or venting of the BC. There are three possible methods of inflation:

a) from a direct feed – this is not the fastest means of inflation but is the most controllable. If the casualty is very negatively buoyant, there will be a short delay between the commencement of inflation and the start of upward motion. For a neutrally buoyant casualty the delay is minimal. Many direct feeds are attached to the BC's mouthpiece, which allows one-handed control of both inflation and deflation without changing grip.
b) from a dedicated air cylinder – this provides the fastest means of inflation and is independent of the casualty's air supply. It is, however, more difficult to control because of the rapid inflation and the necessity to move the hand to a separate control for venting.
c) from a CO_2 cylinder – these are sized to fill the buoyancy device at the surface only and are a 'one-shot' device. The buoyancy provided decreases with depth.

For a drysuited casualty the options available are to use either the suit itself or the casualty's BC. As the drysuit will already contain some air for relieving suit squeeze or for buoyancy adjustment the use of this same device for a controlled buoyant lift provides a simple technique. A grip on the drysuit direct feed connection will provide not only a positive contact but will also allow the same hand to control suit inflation. The other hand is then free to control the venting of the suit by dump valve, cuff or neck seal and to control the rescuer's own buoyancy.

If the casualty's BC is used instead, the air in the BC will have to be vented to counteract not only its own expansion but also that of the air in the suit and, in the case of a neoprene drysuit, that of the suit material. The casualty will not have been raised very far, therefore before all the air is vented from the BC and the rescuer will be forced to change technique in mid-lift to vent from the drysuit. While this technique will work it is unnecessarily complicated.

Some drysuits and BCs are fitted with automatic dump valves. As mentioned in the section on Self-Rescue, these may need to be screwed shut slightly to prevent any air introduced for positive buoyancy being vented straight out again. The characteristics of these dump valves can vary markedly and they only allow for expansion of the air and not, in the case of a neoprene-suited casualty, the expansion of the material. The rescuer should, therefore, also be prepared to use an additional means of venting to maintain a controlled ascent.

The controlled buoyant lift allows the rescuer to control the rate of ascent to a speed suitable to the severity of the situation. For a breathing casualty there is no immediate danger and hence the ascent should be performed at the normal rate. For a non-breathing casualty time is of the essence and a more rapid ascent is required. Paradoxically, it is the rescuer who is most at risk from such an ascent. The casualty, being unconscious, will be fully relaxed and consequently any expanding air in his lungs will vent naturally. The rescuer, due to the stress of the situation, will probably be breathing more heavily than normal and in a rapid ascent is therefore more at risk from a burst lung. Because of the severity of this condition, the urgency to get the casualty to the surface should be tempered with consideration for the rescuer's own safety. Longer-term consequences of a rapid ascent which are less life threatening, such as decompression illness, should not be allowed to detract from the urgency of the situation.

For training purposes where no life is in danger, subjecting the rescuer and 'casualty' to any risk is unacceptable. All practice rescues should therefore be carried out at not more than the normal rate of ascent.

During the ascent the rescuer should ensure that the casualty remains positively buoyant. Should the rescuer and casualty become separated for any reason the casualty will then continue ascending to the surface, albeit now in an uncontrolled manner, where normally he can subsequently be found. Should the separation occur with a negatively buoyant casualty, the casualty will sink, possibly to be lost.

Once at the surface, the casualty's BC should be inflated to provide substantial positive buoyancy. This should be done even if the drysuit has been used to effect

Figure 88 The rescuer controls casualty's inflation/deflation valve

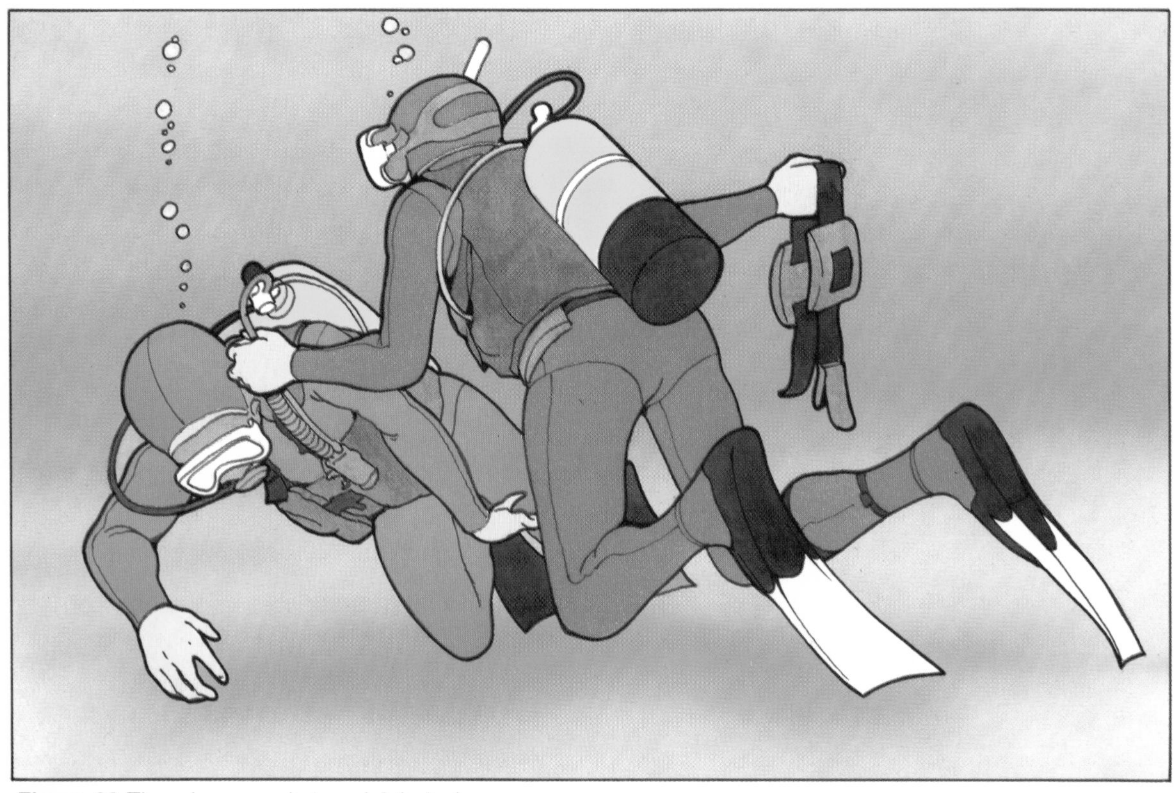

Figure 89 Throwing casualty's weightbelt clear

the lift as the drysuit cannot be relied upon to float the casualty face upwards. Also, should the casualty require AV, disturbance of the neck seal during neck extension may cause deflation of the suit.

Inflation of the BC until it is hard may interfere with adequate neck extension during AV. Venting a small amount of air will allow the BC to 'give' while not significantly affecting buoyancy.

Releasing the Casualty's Weightbelt

Should none of the casualty's buoyancy devices be usable, it may be possible to achieve a measure of positive buoyancy by ditching the casualty's weightbelt. Once unbuckled the weightbelt should not be allowed to just drop, but should be pulled clear of the casualty to prevent snagging on other equipment.

A drysuited casualty will normally be carrying a heavy weightbelt and hence a substantial amount of positive buoyancy will be achieved, resulting in a very rapid ascent. A wetsuited casualty on the other hand, particularly in fresh water, will be carrying considerably less weight and releasing his weightbelt may be insufficient to achieve any positive buoyancy. Depending on the equipment worn by the casualty, the depth, and how accurately the casualty had trimmed his buoyancy to

Figure 90 A controlled buoyant lift using casualty's BC

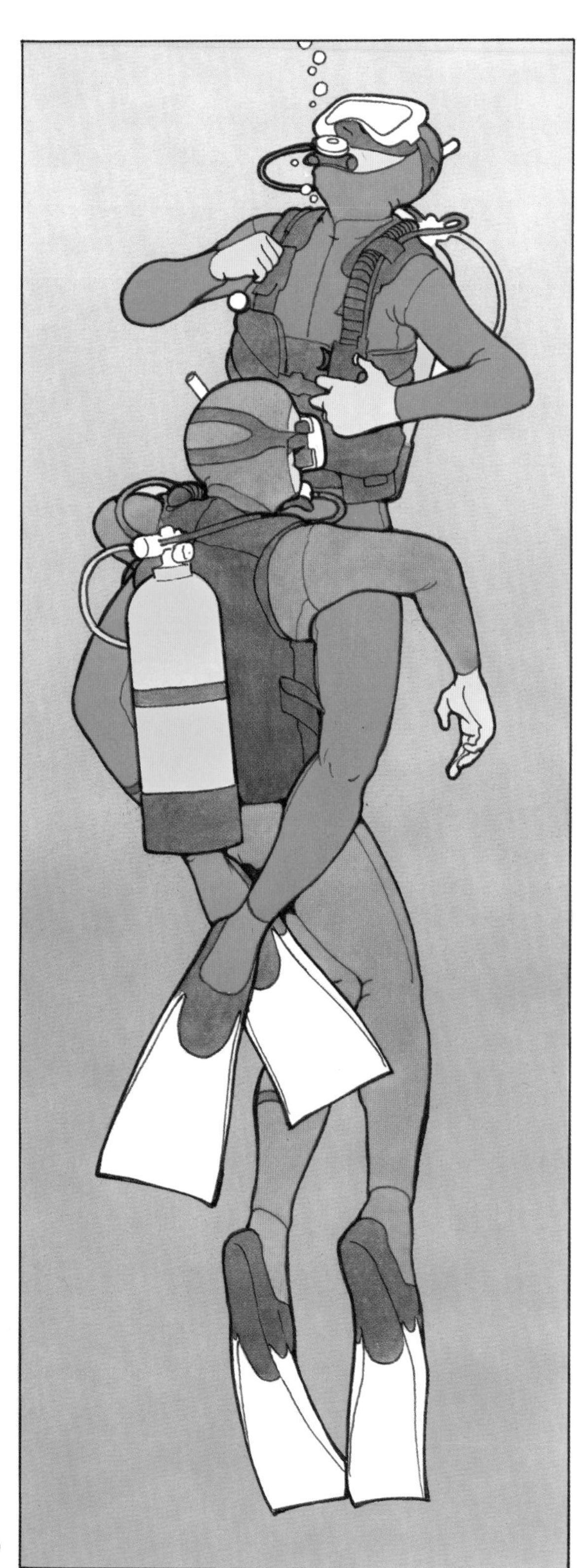
Figure 91

neutral before the incident, the reality of the situation may lie anywhere within these extremes.

The rescuer must, therefore, be prepared to cope with this variability either by dumping some of his own buoyancy and spreading his legs and fins to provide drag to slow the ascent, or by using some other means of achieving positive buoyancy.

The Controlled Buoyant Lift Using the Rescuer's Buoyancy Device

If it is not possible to achieve positive buoyancy for the casualty by any means, the rescuer can use his own buoyancy device to lift the casualty. As this will mean the rescuer being more buoyant than the casualty, a very secure hold on the casualty will be required. Should a separation occur during the ascent, by the time the rescuer has arrested his consequent rapid ascent, the heavy casualty will, in all but very clear water, probably have sunk out of sight.

As the rescuer will possibly require both hands free to control his buoyancy, wrapping his legs around the casualty's waist can effect a secure grip. This grip is, however, more difficult if the casualty is wearing a twin-set or other bulky gear (see Figure 91).

Once the grip is established the technique becomes the same as described in Self-Rescue. During the ascent the casualty may subsequently become buoyant due to air or suit material expansion and hence the rescuer should be prepared to change his technique to suit the changing situation.

The Swimming Lift

If all else fails the rescuer will have no recourse other than a swimming lift. This is a very strenuous task and experience has shown that the extreme effort involved could prejudice the safety of the rescuer. The consequent dramatic increase in the rescuer's air consumption may also prove critical if his remaining air supply is limited. This type of rescue should, therefore, only be attempted as a last resort!

In order to prevent the casualty from interfering with the rescuer's fin strokes, he should be gripped in such a way that he is held high relative to the rescuer. The further below the casualty the rescuer can position himself the less tendency there will be for the weight and drag of the casualty to turn their motion into a curve away from a vertical ascent. Particularly if the casualty is wearing a heavy cylinder, this can be minimized by holding the casualty from behind, where a large part of the weight will be.

The Uncontrolled Ascent

The controlled lift provides the most efficient means of raising a casualty to the surface without putting the rescuer at undue risk. It is important, however, that should the situation get out of hand it will still result in an ascent, albeit uncontrolled. At the surface the consequences of the uncontrolled ascent can at least be treated. If the failure results in a descent then little can be done and the outcome will almost certainly be fatal.

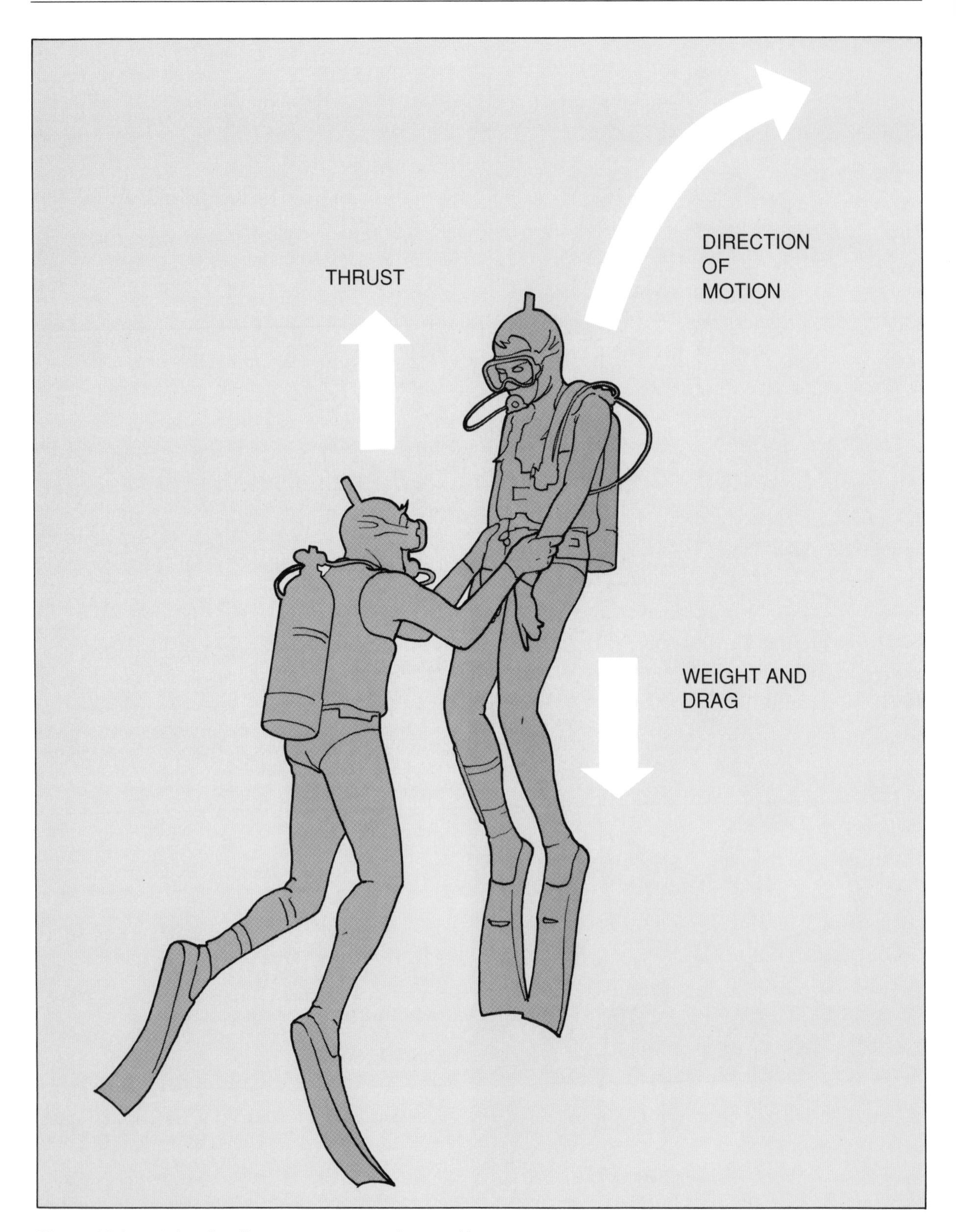

Figure 92 In a swimming lift rescuer assumes low position

Figure 93 A swimming lift with grip on cylinder gives most vertical line of ascent

Landing

After completion of a tow it will be necessary for the casualty to be removed from the water. The actual technique that may be used depends upon several contributory factors: the relative size of the casualty and rescuer, the site of landing, assistance available and the conditions at the site of landing. It is important to construct a plan that will include considerations of landing when the rescue is assessed if these difficulties are to be minimized. Consider the different methods of landing available:

Landing into a Boat

The techniques of landing into a boat will depend upon the size of the boat as well as the assistance or aids available. Consideration must be given as to the stability of the boat before the landing commences if the boat is not to be capsized during the landing. The methods available are primarily a straight-arm lift or using ropes or similar aids for an assisted lift.

Figure 94 The conscious casualty will usually require assistance to leave the water. Both rescuer and boat crew should assist

Figure 95 Remove casualty from the water by the swiftest means, using help when available

Straight-arm Lift

It is important during this type of landing that careful consideration is given to the safety of the casualty's head if injury is to be prevented. Consideration should also be given to the safety of the rescuer, since incautious lifting could damage the rescuer's back. Prior to lifting the casualty, equipment should be removed or ditched to minimize the difficulties of landing, and the rescuer should also remove his own equipment. If assistance is available it can be used to help with the removal of equipment and the maintenance of artificial ventilation (if needed) during the landing (see Figure 95).

1. The casualty's hands are placed together above the water and held by the rescuer.
2. The rescuer can then climb out whilst maintaining a hold on the casualty's hands to prevent him slipping back.
3. The rescuer grips the casualty's wrists or elbows as most convenient.
4. The casualty is raised to waist level and folded over the side by the rescuer stepping back. Care should be exercised to avoid raising higher than waist level if injury is to be avoided (see Figure 97).
5. As the casualty is lowered to the ground a hand placed in the middle of the back will prevent the casualty slipping back. Care is taken to protect the face and head during this process.
6. The casualty can be turned onto his back for further artificial ventilation or placed into the recovery position as required.

This method may be modified by the use of two rescuers and can also be assisted by pulling the thigh once the casualty has been raised. Using this method it is possible to land a casualty into any boat where the freeboard does not exceed the length of the rescuer's arms. In small boats it may be necessary to land over the transom if stability is to be maintained during the landing. Assistance can also be given by using a rope around the casualty's chest under the armpits, but care must be exercised to ensure that breathing obstructions are not generated.

Figure 96 Recovering casualty by parbuckling

Figure 97 Recovering by straight-arm lift

Assisted Lift (Parbuckling)

It is easy to use a rope or other aid such as a sail to assist the lifting of the casualty from the water. The rope needs to be very carefully positioned to avoid slipping around the casualty's neck or to his knees, where it can cause the casualty to fold up. The rope should be positioned by an assistant in the water, whilst tension is maintained from within the boat. The technique gives a mechanical advantage to the lift, thereby reducing the effort required to lift a casualty or allowing heavier casualtys to be lifted (see Figure 96).

1. The rope is carefully positioned around the casualty's chest with the arms inside the bight of rope. The inboard end of the rope should be fastened securely and the free end passed back to the assistant in the boat. The lower strop is positioned at upper thigh level and should be level with the chest rope.
2. The free end of the rope is then pulled, both sections being pulled evenly to maintain a level position. As the casualty rolls up the side of the boat care must be taken to ensure that the rope does not slip and that the lift is even. The head and face should also be carefully protected during the landing.

Using this method it is possible to raise a casualty into a boat over quite large distances, but care must be taken to ensure no slippage occurs. This method can also be used to land a casualty onto a jetty or a quay side where there is no easier access.

Landing onto a Beach

The method of landing that can be used will depend upon conditions as well as the relative size of the casualty and rescuer, but once again will require removal of equipment for the most efficiency.

Sloping or Sandy Beaches

The simplest method of landing someone that does not require actual lifting is to drag them ashore. This is made easier if the rescuer stands behind the casualty and holds the casualty's arms across his abdomen, with the rescuer gripping the casualty's wrists. The weight of the casualty against him will maintain the grip without the need for the rescuer to exert force on the wrists. By simply walking backwards the casualty can be taken ashore, although care is needed that the rescuer does not trip as he walks backwards (see Figure 99).

This technique can also be modified by the rescuer laying the casualty onto his back and grasping the casualty's arms over his shoulders. This will allow the rescuer to observe his route ashore, but does require more effort (see Figure 98).

Rocky or Steep Beaches

The Fireman's lift can be used if the casualty is smaller than the rescuer, but great care must be taken to ensure that the casualty is not pushed underwater during the lift and that the rescuer does not overtax himself (see figures 103 and 105).

1. Lay the casualty on his back in front of you and grasp the wrist nearest to you firmly with your left hand. Place your right hand between the casualty's legs and grasp at about the knee level.
2. Roll the casualty onto your shoulders by ducking yourself under his armpit at the same time as pulling him into position.
3. Stand up and adjust until comfortable

This method can also be used for climbing ladders up the side of quays, but does require care if it is to be successfully performed without injury to casualty or rescuer.

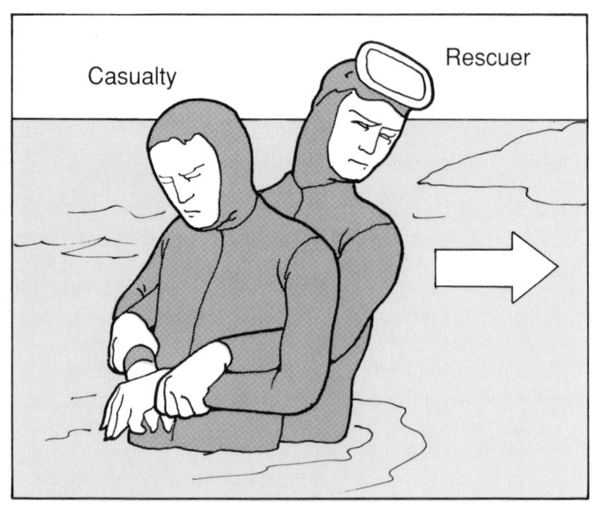

Figure 99 Proper grip for drag lift

Figure 100 Single rescuer drag lift

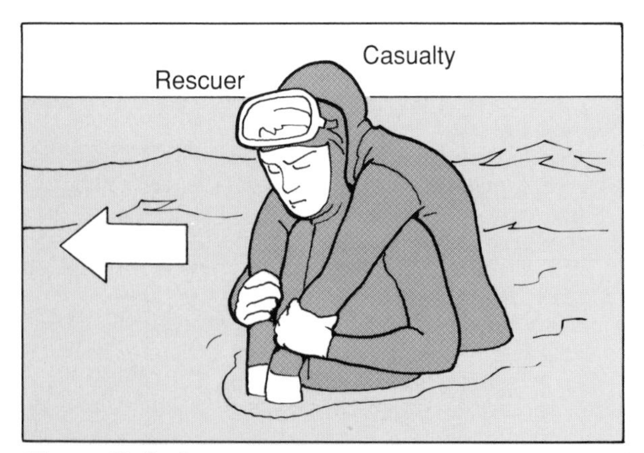

Figure 98 Back carry

Figure 101 Two rescuer drag lift

Figure 102 Supporting a walking casualty

Figure 104 A three rescuer carry

Figure 103 Picking up a casualty for Fireman's lift

Figure 105 Carrying casualty in Fireman's lift

Landing the casualty into a hard boat

Landing a casualty into a large boat is a complex exercise, which will never be the same twice. The infinite variety of configurations of boat and conditions under which a rescue may have to be performed prevent any hard-and-fast technique being proposed. This discussion is therefore intended to acquaint the reader with the factors which need to be considered and to illustrate these with some possible techniques which can be used as a starting point for adapting to specific conditions when the need arises.

The following factors need to be considered:

Freeboard: The higher the freeboard of the boat then, obviously, the greater the height through which the casualty has to be lifted. This poses two problems. Lifting the casualty through a greater height will require more physical effort, and beyond a certain height rescuers on the boat may not be able to reach directly a casualty in the water. This second factor may be further aggravated by obstructions around the gunwale, such as railings. Sharing the effort between as many people as possible will clearly make the lift easier, but there may not always be sufficient access to do this. In addition to rescuers on deck, rescuers stationed below deck level on the diving ladder or lower still if a small cover boat is available can provide further assistance. To make the lift as easy as possible all the casualty's heavy equipment should be jettisoned before the lift commences.

Boat Motion: The motion of the boat influences the rescue in a number of ways. The rescuers will be working from an unstable base. Depending upon the sea state, the motion will range from insignificant to the factor causing most problems in the whole exercise. As boat motion increases the rescuers will need to devote more effort to holding on, just for their own security, and this will inevitably detract from the effort they can apply to lifting a casualty.

As the boat rises and falls the freeboard, and hence the height through which the casualty has to be lifted, also changes. This is complicated by the fact that, as the boat rolls, the gunwale describes an arc above the sea surface alternately moving towards and then away from the casualty in the water. The lift of the casualty therefore needs to be timed to occur as the gunwale approaches its closest point to the casualty.

Protection: To protect both the casualty and any rescuers in the water from being struck by the boat hull or fittings as it rolls, they should remain a couple of metres clear of the hull until preparations for the lift are complete. They should then spend the minimum time necessary alongside the boat for the lift to take place. The casualty also needs to be protected to prevent impact with the hull of the boat during the lift.

Co-ordination: Because a number of rescuers need to be involved in this situation it is essential that all the efforts are co-ordinated to ensure that the lift is performed quickly and efficiently. This is important, not only for the safety of all concerned but, if the casualty requires artificial ventilation, to minimize any interruption.

Figure 106

Figure 107 Lifting casualty from cover boat to hard boat

Many hard boats operate with a small inflatable as a cover boat. Should an incident occur, the cover boat will make the initial pick-up, but the problem of transferring the casualty to the hard boat still remains. With the cover boat providing a base to work from, the casualty can be lifted initially by two rescuers in the cover boat, each grasping the casualty under the armpits. The casualty can then be raised sufficiently for two more rescuers reaching down from the hard boat to continue the lift. The rescuers will find the casualty easier to grip if he is facing outboard from the hard boat, and this has the added advantage of protecting the casualty's head from its natural tendency to fall forward and strike the hull.

Timing of the lift is crucial, so the rescuers in the cover boat must raise the casualty to coincide with the maximum rise of the cover boat relative to the hard boat. Rescuers in the hard boat will then have the least distance to reach down, but should be prepared to take the casualty's weight quickly as the cover boat drops away again. If this is not adequately co-ordinated there is a high risk of the casualty being dropped back into the cover boat or, worse still, into the water between the two boats with all the additional injury that he may sustain.

When no cover boat is present the lift will have to be made all the way from the water. With all the casualty's heavy equipment jettisoned he should be brought alongside the boat, facing outboard and close to one side of the diving ladder. A rope from the deck, passed under the casualty's armpits can provide the main lifting force. Another rescuer stationed on the diving ladder can provide extra lift and protection for the casualty to prevent him swinging as the boat rolls. Once rescuers on deck can reach down to the casualty the rope is dispensed with.

On smaller hard boats, of the day boat variety, where the freeboard is much lower and the deck is recessed inside the hull close to the waterline, consideration should be given to lifting the casualty facing inboard. The rescuers will be standing at a level somewhat below the gunwale, over which they will have to lift the casualty. Taking advantage of the natural 'hinge' in the casualty's middle will allow him to be folded over the gunwale with much less effort than by trying to bend him backwards. While this requires that more attention be paid to protecting the casualty's head, it makes the lift considerably easier and prevents damage to the casualty's back. In larger hard boats where the rescuers' feet are roughly at the level of the gunwale they can lift high enough for this not to be a problem.

Figure 108 Boat crew assisting casualty from water

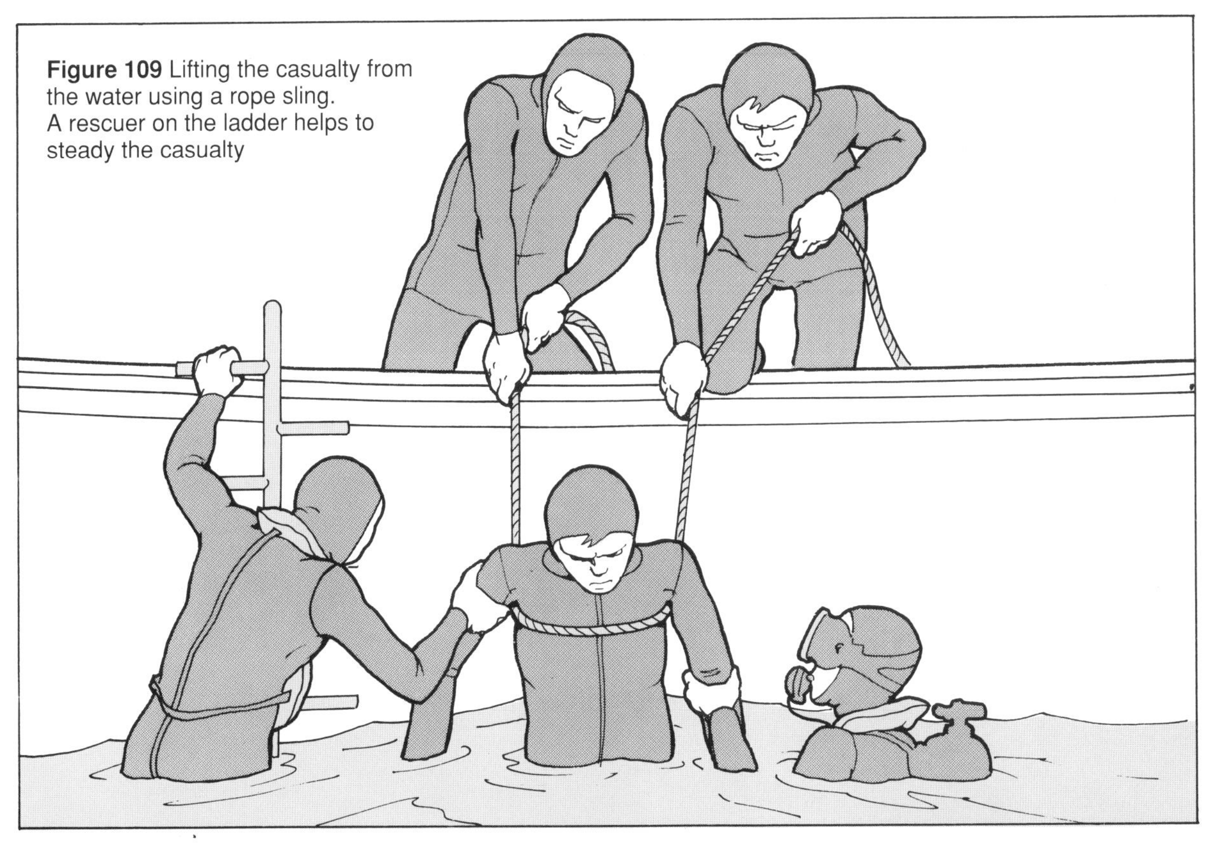

Figure 109 Lifting the casualty from the water using a rope sling. A rescuer on the ladder helps to steady the casualty

Figure 110 A sling will also be useful with a lower freeboard boat

Figure 111

Rescue Management

Rescue Management

The management of a rescue should begin without delay. Although medical and recompression facilities may be close to the incident, this will not assist an injured diver in an inflatable boat out at sea. The rescuer's task is to ensure that the injured diver reaches those facilities in a reasonable condition, so that further professional medical care can be given. The rescuer must, therefore, possess a reasonable level of knowledge and skill in three main areas:

1. Rescue – The removal of the casualty from danger
2. First aid – Temporary care and relief
3. Accident management – On-site organization and control

Although a rescuer should be capable of operating in each of these three main areas, an assistant who is qualified to take over the first aid, or oxygen administration, will prove invaluable to anyone trying to manage a rescue situation. In the ideal situation, having a number of capable assistants at each stage of the rescue enables the person in charge to adopt the role of co-ordinator, thus giving him the flexibility to employ his resources to the maximum benefit of the casualty. There are, of course, circumstances in which the rescuer will have to perform all three roles. However, in practice most emergencies are a mixture of both extremes.

Because few divers perform more than two rescues in a lifetime, it is difficult to recognize the signs that something is happening, or has happened. Accident casualties usually die in the presence of others who could have assisted if only they had recognized the problem. However, most successful rescues involve certain general principles, which if applied correctly, make a rescue more likely to succeed. One thing is certain, however, no two rescues will ever be the same!

Figure 113 Reassure and comfort the casualty

Figure 112 Remove the casualty from danger

Priorities of a rescue

To be effective the rescuer will need to *assess* the situation, *plan* a course of action, and then *act* quickly.

ASSESS
PLAN
ACTION

Assessment

Assessment is a continuous process of evaluation of all the relevant factors throughout a rescue situation. This process can begin even before an actual emergency occurs. Experienced divers will assess the conditions for diving upon arriving at the boat or dive site, and can often recognize circumstances which could lead to an incident. The assessment of the diving party is another important factor. Checking the recent diving experience and qualifications of the diving party may reveal important information. Knowing that there are divers with lifesaving skills, doctors, paramedics, or other specialists may help you to pre-plan a possible course of action. Watching while divers prepare their equipment for diving can also give a good indication as to their experience and ability. Are they nervous, unfamiliar with their equipment, under stress, out of physical condition?

Just asking a person how they feel about the forthcoming dive can often prevent incidents. If concern is expressed, then the dive can be postponed; if not, then at least the rescuer is aware of a potential problem.

While diving operations are in progress, other signs of an impending emergency can assist the rescuer in anticipation.

Divers surfacing a long way from the boat.
Divers surfacing downstream from the boat in a tidal flow.
Divers surfacing near rocks in a heavy swell.
Lone diver on the surface.
Deteriorating weather conditions.
Other craft in the diving area.
Unreliable outboard engine.

Obviously, when the possibility of an incident can be anticipated, rescue procedures can be planned, thus saving valuable time if things do go wrong.

Figure 114 Deteriorating weather conditions

Figure 115 Diver signalling distress

Plan

A common and often serious mistake made by the would-be rescuer is to leap into the water without stopping to think. The first consideration in any rescue attempt should be self-preservation. If the rescuer gets into difficulties, he can be of little assistance to the casualty. The next most important consideration should be whether there is an alternative solution to effect the rescue, which does not require getting wet. Can the diver be reached with a paddle, a rope, or is there a buoyant aid that can be thrown? Is there a boat in the area, or would it be quicker to alert the emergency services?

In a diving emergency situation, where a diver has surfaced and needs immediate assistance, the rescuer needs to be able to respond quickly, but at the same time evaluate the situation and formulate a plan of action. A few moments is all the time an experienced rescuer will need to determine a plan of action. This evaluation may include some or all, of the following:

What is the divers condition? Is he conscious or unconscious?

Is the diver floating or sinking?
How far away is the diver?
What are the sea conditions?
Is there another boat available?
Is his buddy able to assist?
How can help be summoned?
What emergency equipment is available?
Is anyone else responding?
Is there more than one diver in trouble?

Types of Diving Incident

Strictly speaking the term decompression illness applies to all injuries sustained due to the reduction of pressure to which a diver has been subjected. In practice, however, the term tends to be used to refer to those conditions which arise by absorbed gas causing bubbles to form in the blood or tissues or from bubbles introduced directly into the blood due to lung damage (air embolism). Other forms of lung pressure damage (barotrauma) which cause gas bubbles to escape between the body organs are referred to as burst lung.

Irrespective of the cause, all forms of decompression illness and burst lung have certain similarities that make the on-site first aid the same for all.

Decompression Illness due to Absorbed Gas

The gases dissolved in the human body are normally in equilibrium with those in the atmosphere. When a diver descends the increased ambient pressure to which he is subjected disturbs this equilibrium and more gas is absorbed into the body, via the lungs, as nature tries to restore the balance. The amount of extra gas that is absorbed depends on the depth (pressure) to which the diver descends and the time for which he remains there. As the pressure is reduced (decompression) during the ascent at the end of the dive the reverse happens as the extra gas absorbed is eliminated from the body to re-establish the surface equilibrium.

If the ascent is performed too rapidly the gas cannot be eliminated from the body through the lungs quickly enough and bubbles form in the blood and tissues. By proper dive planning using appropriate 'decompression tables' this condition can be avoided and ascents made in safety. In almost any decompression, bubbles will in fact form in the blood stream, but they are normally so small that they are of no consequence. If, either through error or accident, an ascent is made too rapidly, the size of the bubbles formed becomes sufficient to interfere with bodily functions in a number of ways collectively known as 'decompression illness'. The bubbles may form in either the blood or the tissues.

As the pressure reduces, the bubbles increase in size until they obstruct the blood vessel, reducing the blood flow to tissues downstream of the blockage. As with air embolism, the effect this has depends very much on which tissues are suffering the reduced blood (and hence oxygen) supply.

Within the tissues the formation of bubbles can have two effects. The physical presence of the bubbles exerts pressure on the surrounding tissues causing damage. Where bubbles occur close to a blood vessel the pressure exerted by the bubbles on the wall of the blood vessel can cause a restriction in the size of the blood vessel, thus interfering with the blood supply to down-stream tissues.

Prevention

The risk of decompression illness can be minimized only by conscientious dive planning and the correct use of recognized decompression tables in conjunction with conscientious monitoring of depth and time. To avoid misleading information due to instrument error, regular depth gauge calibration checks are recommended. Modern decompression computers will perform the above functions automatically, but regular calibration checks of the depth read-out are still essential.

The risk of a diver being subject to decompression illness is increased by anything which adversely affects the diver's physical condition such as obesity, seasickness, dehydration, 'the morning after the night before', etc. Continual diving over a period of days can also result in a cumulative effect causing decompression illness to occur as a result of dives carried out within the parameters of decompression tables.

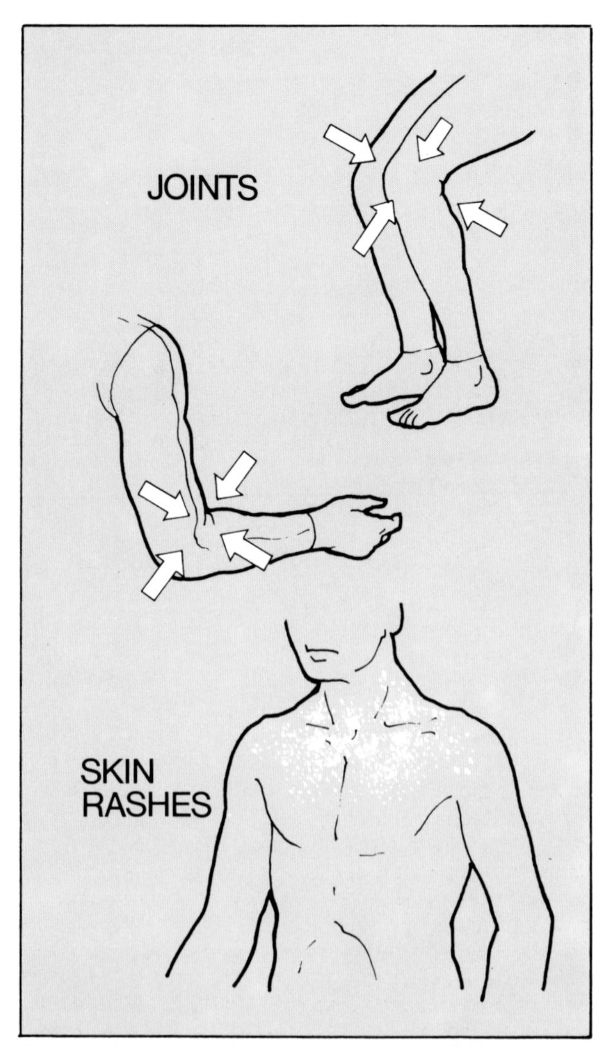

Figure 116 Symptoms of decompression illness

Symptoms due to Absorbed Gas

Symptoms of decompression illness occur anywhere from a few minutes after a dive to several hours afterwards. Generally the shorter interval is associated with short deep dives and the longer interval with longer shallower dives. Experience has shown that 50 per cent of cases will show symptoms in less than one hour, 90 per cent within six hours and very few in excess of twenty-four hours. Symptoms occurring after forty-eight hours or more are unlikely to be due to decompression illness. Symptoms are usually classified by the body tissue affected.

Joints

Pains may occur in any joints, but are most common in the large joints of the shoulders and the knees. Very mild pains are called 'niggles'. In more serious cases the severity of the pain usually increases over a period of twelve to twenty-four hours, becoming a deep throbbing pain with occasional sharp exacerbations. The characteristic flexing of the joint to ease the pain is what gave the name of the 'bends' to decompression illness.

Skin

Itches, rashes, irritation and mottling are in themselves less serious forms of decompression illness and are often left to resolve naturally. Until proved otherwise, however, they should be heeded as possible warnings of more serious conditions.

Nervous System

Depending upon which part of the nervous system is affected a wide range of symptoms can occur including, but not necessarily limited to, the following:

Numbness — balance disturbances
tingling — nausea
weakness — confusion
paralysis — convulsions
visual disturbances — unconsciousness
loss of bladder/bowel control

Symptoms of this kind can by their nature dramatically affect a diver's judgement. The companions of a diver suspected of suffering from decompression illness should therefore be wary of the divers assessment of his own condition. It is not unknown for seriously affected divers to refuse to believe that anything is wrong with them.

Heart and Lungs

After long deep dives, inadequate decompression can result in the formation of considerable quantities of bubbles in the blood stream. The capillary bed of the lungs acts as a fine filter generally preventing these bubbles reaching the arterial circulation. The bubbles collect as froth in the lung capillaries and in the right side of the heart, interfering with the blood circulation and resulting in low blood pressure. This condition, aptly named the 'chokes', is an extremely critical condition and if not treated promptly can be fatal. It usually occurs shortly after surfacing and is characterized by the affected diver complaining of being short of breath, having a tight feeling across the chest, and appearing cyanosed and suffering severe shock. In extreme cases some bubbles may find their way through the lung capillaries and into the arterial circulation where their further effect will be the same as for air embolism.

A further mechanism may also allow bubbles to pass into the arterial circulation. In the human heart, at birth there is an opening, which allows blood to pass directly from the right to the left sides of the heart (i.e., from the venous to the arterial circulations). Shortly after birth this opening is sealed by a flap of tissue. In some people, and under certain conditions, this flap forms a less than perfect seal (known as patent foramen ovale, PFO). Should nitrogen bubbles form in the blood, it is possible that they may pass, via a PFO, from the venous to arterial circulations, again producing effects similar to those of air embolism.

Gut

Relatively infrequently affected by decompression illness but manifest by abdominal pain, vomiting or diarrhoea. The latter two may show evidence of bleeding.

Bones

Decompression illness of the bones, known as bone necrosis, is a long-term problem, which can affect professional divers or caisson workers. It is extremely rare in amateur divers where it is usually the result of persistent inadequate decompression. Bone necrosis is a disruption of the blood supply to the bone, resulting in the death of areas of bone structure. This is particularly a problem where the dead bone forms part of a joint and results in a painful deformity of the joint structure.

Decompression Illness due to Air Embolism

Damage to the capillary wall of the alveoli allows bubbles of air (emboli) to enter the blood stream directly. From the lungs this blood passes via the heart into the arterial circulation. Any air bubbles are therefore rapidly distributed around the body until they encounter blood vessels, which are too small to permit their passage, where they form an obstruction. Blood flow to tissues downstream of the blockage is thus impeded, producing a wide variety of symptoms dependent upon the location and severity of the obstruction.

Typical symptoms are:

giddiness
nausea
numbness
paralysis
visual disturbances
unequal-size pupils
mottling of the tongue
changes of mood
respiratory difficulty
heart failure
death

The common feature of all symptoms of air embolism is that the onset is extremely rapid after reaching the surface, typically within seconds.

Burst Lung

This condition results from inadequate ventilation of air from the lungs during the ascent causing damage to the tissues of the lungs. The consequences can be extremely life threatening and require prompt action.

Inadequate ventilation of the lungs can arise from a number of causes. Panic due to an emergency may cause a diver to forget his training and to revert to his natural instinct to hold his breath while making for the surface. This may not be deliberate, as a possible symptom of panic is a spasm of the epiglottis, which will prevent the diver from exhaling. Even if the diver is not in panic but is making a free or a rapid buoyant ascent, he may fail to breathe out at a fast enough rate to counteract the expansion of air within his lungs. Because the lung's 'full-to-bursting' signal (the Hering-Breuer reflex) is extremely weak in man, the diver would feel no physical sensation of the damage being caused to the lungs. It is for this reason that the BSAC does not recommend the practising of free or rapid buoyant ascents.

Local trapping of air within the lungs may occur due to inflammation or infection. While this may limit the area of the lungs subjected to physical damage, it will not limit the severity of the condition.

As the air in the lungs expands during the ascent the lungs expand to their maximum volume. Further ascent then causes the tissues of the lungs to be stretched until eventually tearing occurs. Once the lungs are full, an ascent of as little as one metre is all that is required to cause damage.

Prevention

On entry to the sport, divers are recommended to undergo a thorough sport diving medical examination. Lung conditions, which may preclude participation in the sport, will thus be identified. Transient medical conditions affecting the lungs should preclude diving until the condition has cleared.

While diving, a normal breathing rhythm should be maintained at all times, particularly during the ascent. Should a situation requiring a faster than normal ascent occur, remain calm and concentrate on controlling your exhalation. (see Pages 56–61 – Self-Rescue).

Symptoms

Depending upon the physical damage sustained by the lungs, one or more of two different types of burst lung may occur.

Spontaneous Pneumothorax

Damage is caused to the lung sac; air can escape into the space between the lung sac and the chest wall. As this air expands the lung sac is caused to collapse reducing the area of the lungs available for gas exchange.

Typical symptoms are:
shortness of breath
coughing up bloody sputum
pain on breathing
swollen appearance of the chest cage
cyanosis

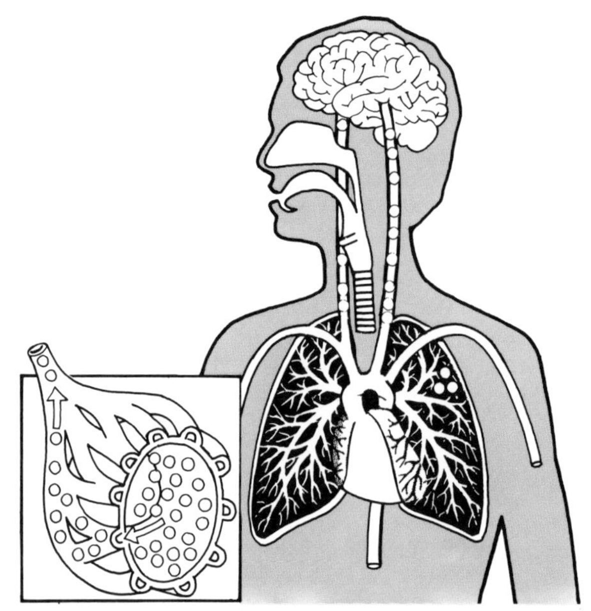

Figure 117 Air Embolism

Interstitial Emphysema

Air escaping inwards from the lungs can work its way through the organs between and above the lungs. As the air expands pressure may be exerted on various organs including the airways and blood vessels in the neck, and the heart and lungs in the chest.

Typical symptoms are:
shortness of breath
difficulty in swallowing
swollen base of the neck
changes in voice quality/tone

While air embolism can occur on its own, it is unusual for either of the above two conditions to occur without simultaneous air embolism. The symptoms exhibited could, therefore, be a combination of any of the above.

First Aid for Decompression Illness and Burst Lung

As a general rule any unusual symptoms following diving should be assumed to be due to decompression illness or burst lung.

The on-site treatment for decompression illness is identical to that for burst lung. Subsequent therapeutic treatment may differ considerably, but that is the concern of specialist medical personnel not the diver on the spot.

The casualty should be laid flat on his back, kept warm and treated for shock, 100 per cent oxygen should be administered at the earliest opportunity. The elimination of nitrogen from the inspired gas maximizes the nitrogen pressure gradient between the inspired gas and the nitrogen in the blood, hence speeding the elimination of nitrogen from the casualty's body.

100 per cent oxygen will also increase the partial pressure of the oxygen in the blood and hence help to offset any reduction in blood flow due to capillary obstruction by bubbles. In cases of lung sac collapse the

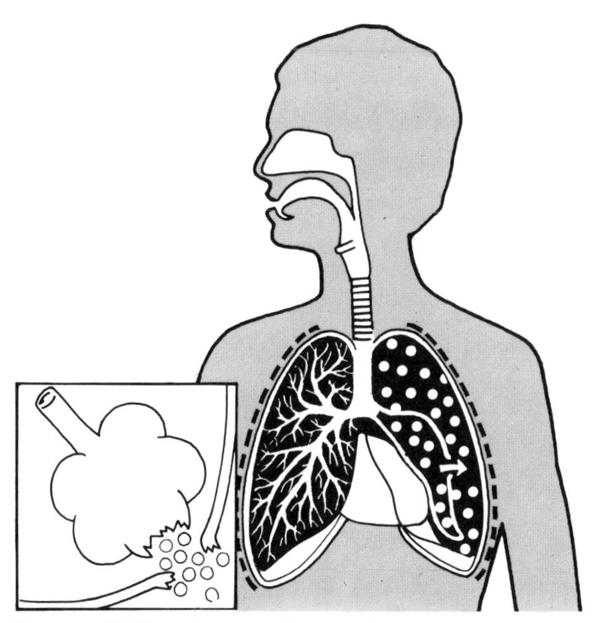

Figure 118 Spontaneous Pneumothorax

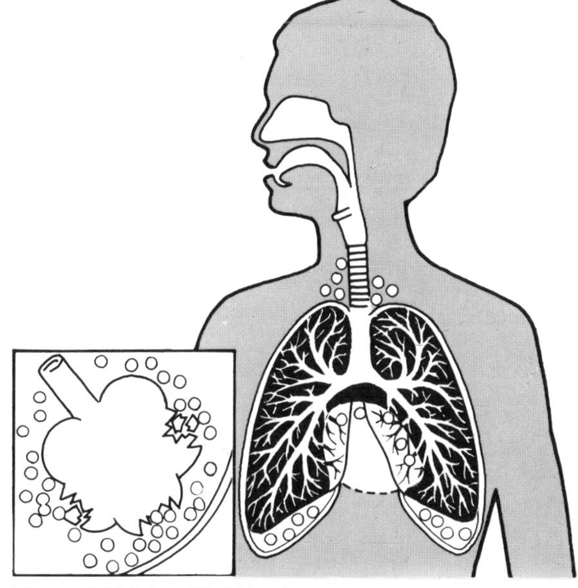

Figure 119 Interstitial Emphysema

greater inspired partial pressure of oxygen will increase oxygen exchange across the remaining undamaged areas of the lung.

The casualty's breathing and pulse should be continuously monitored. If either fails, artificial ventilation, preferably oxygen enriched, and Chest Compressions should be performed as necessary, while arrangements are made for the urgent evacuation of the casualty to medical attention.

The only effective treatment for air embolism is recompression at the earliest opportunity in a recompression chamber. Unless there is one on site, arrangements should be made for the urgent transport of the casualty to an appropriate facility.

Without in any way interfering with the evacuation of the casualty to medical assistance, perform an evaluation of the casualty's symptoms commencing at the earliest opportunity (see Pages 22–23).

Whether suffering from decompression illness or burst lung, the casualty will be a very sick person and severely incapacitated. Consequently in-water recompression should not be contemplated.

The casualty's buddy will have been exposed to the same dive profile as the casualty and consequently should be treated as a potential casualty. The buddy should therefore be closely monitored and, as a precaution, should accompany the casualty to a suitable medical facility unless his recent diving history is known to be sufficiently different.

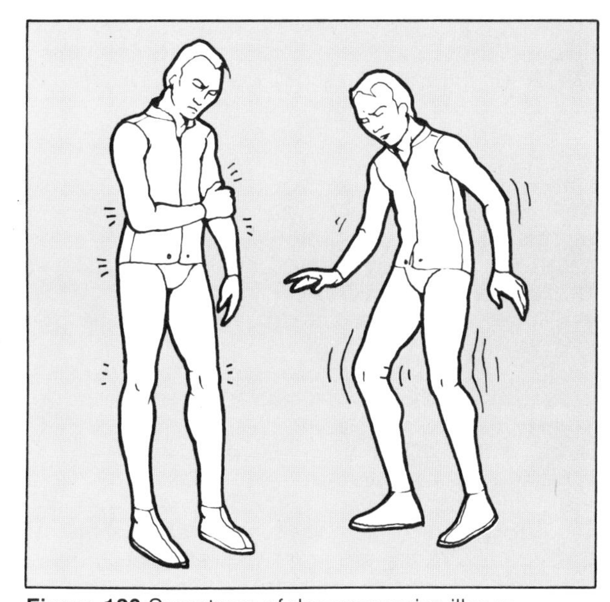

Figure 120 Symptoms of decompression illness

Nitrogen Narcosis

As a diver descends, the pressure around him increases and the partial pressure of the nitrogen in the air he breathes also increases. Normally regarded as inert, in that it is not used in metabolism, the increased amount of nitrogen dissolving in the brain cells begins to affect the diver's mental processes. At first the effects are so mild they are not noticeable, but as depth increases the effects of nitrogen narcosis also increase to alter the diver's awareness of his situation and his behaviour.

As the effects take hold, simple tasks become awkward. Gauges are misread, signals are misinterpreted. As the effects increase the diver makes irrational decisions and becomes reckless, while at the same time becoming increasingly convinced that he is diving in a safe and conscientious manner. Nitrogen narcosis thus puts the diver at increased risk of an accident while decreasing his ability to cope with the situation.

The mechanism by which nitrogen narcosis works is the subject of much speculation, but few hard facts. What is certain, however, is that while there is considerable variation between divers, no diver is immune from it. Few amateur divers will not show significant degradation of their performance at depths of 30 metres or so, some at even shallower depths. At 50 metres the effects are quite significant. A diver claiming that he has not been affected at these depths is merely proving, quite conclusively, that he has been!

Prevention

It is not possible for the amateur diver to prevent nitrogen narcosis, but being conscious of it and able to recognize its effects allows the problem to be minimized. Work-up dives to progressively deeper depths allow some acclimatization, but amateur divers are unlikely to be able to dive to sufficient depths often enough to sustain it.

The diver must recognize that nitrogen narcosis will place limits on the depth to which he can safely dive. These limits are not precise rules, which can be laid down, and only the individual can decide what is sensible for himself. In doing so it should be borne in mind that the dangers associated with nitrogen narcosis become so significant at a depth of 50 metres that, in the United Kingdom, regulations governing commercial diving prohibit professional divers from exceeding this depth while breathing air. For dives deeper than 50 metres special gas mixtures are used which have a less narcotic effect than air. The equipment, logistics and control necessary for this sort of diving are far beyond the means and capabilities of most amateur divers.

To minimize the effects of nitrogen narcosis, all major decisions affecting the dive should be taken on the surface before the dive commences and then the dive plan followed conscientiously. Trying to use a decompression table on a slate at depth vastly increases the chances of the wrong dive profile being followed.

Symptoms

The accompanying table shows the main symptoms likely to be experienced by a diver suffering from nitrogen narcosis. It should be appreciated that these were determined during dry compressions in a chamber and would become apparent at shallower depths when underwater.

The symptoms are experienced a very short time after reaching depth and are purely a function of depth. Remaining at a given depth for a period of time does not cause them to get worse. A rapid descent is likely to aggravate the symptoms. There is considerable variation in the symptoms experienced by different divers at the same depth and likewise in those experienced by a particular diver on different occasions at the same depth.

An individual's susceptibility is increased by cold, poor visibility, exertion, anxiety, alcohol, sedative drugs (including anti-seasickness drugs) and inexperience. On the other hand a strong motivation and acclimatization will increase the individual's tolerance.

Treatment

Symptoms of nitrogen narcosis will improve almost immediately if the diver ascends to a shallower depth. There are normally no after-effects although it is not unusual for the diver to subsequently have no recollection of either the symptoms or his actions while affected.

During deep dives each diver should monitor not only himself for symptoms of nitrogen narcosis, but should also observe his buddy. If a diver behaves unusually or fails to respond to signals, his buddy should be prepared to assist him to ascend. Caution should be exercised in how this is done as the narcotized diver may misinterpret his buddy's actions. Remaining behind or to one side of the affected diver should allow assistance to be given while enabling the rescuer to distance himself if necessary.

EFFECTS OF NITROGEN NARCOSIS

DEPTH	MOOD	INTELLECTUAL FUNCTION	RESPONSE TO STIMULI	BALANCE AND COORDINATION
10M				
20	Mild euphoria	Mildly impaired reasoning		
30			Delayed response to stimuli	Little impairment
40	Over confidence and laughter	Calculation errors		
50				
60	Hysterical laughter	Confusion and drowsiness	Severe delay in response	Dizziness and impaired dexterity
70				
80	Hallucinations	Stupor	Unconsciousness	Severe impairment of dexterity
90				
90+		DEATH		

Carbon Monoxide Poisoning

Carbon monoxide is a colourless, odourless and tasteless gas, a product of the incomplete combustion of carbon containing compounds. For divers, the most common source is the exhaust of internal combustion engines (both petrol and diesel).

Under normal circumstances haemoglobin combines with oxygen to form oxy-haemoglobin and this unstable compound provides the body's major means of oxygen transport to the tissues as it circulates in the blood stream, breaking down in the body tissues to supply them with essential oxygen. However, when carbon monoxide is inhaled it combines with the haemoglobin to form carboxy-haemoglobin. Carboxy-haemoglobin is some 200 times more stable than oxy-haemoglobin and therefore locks up the haemoglobin, making it unavailable for oxygen transport. Hypoxia results. The severity of the hypoxia will depend on the partial pressure of the inhaled carbon monoxide.

The body can tolerate only a certain partial pressure of carbon monoxide at the surface. As the diver descends the pressure of his inhaled air, and hence the partial pressure of any carbon monoxide present, increases to match the water pressure. At depth the oxygen-carrying capacity of the blood is increased by extra oxygen dissolving in the blood plasma and this can to some extent offset the lowering caused by the raised level of carboxy-haemoglobin. Even so, as the diver ascends and is thus subject to falling pressures, the oxygen-carrying capacity falls and may reach a level at which consciousness cannot be sustained.

Once carboxy-haemoglobin is formed its relative stability means that it takes a considerable time to break down and for the carbon monoxide to be eliminated from the body.

Prevention

The most common cause of carbon monoxide poisoning is contaminated air. This occurs when cylinders are filled by a badly sited or maintained compressor.

Compressor operators should ensure that the air inlet to the compressor is situated where it is drawing in clean air. If the compressor is a portable unit the most obvious source of contaminants is the exhaust of the compressor unit motor, but other adjacent sources of contamination, such as vehicle or boat engines, should not be overlooked. A badly maintained compressor can be a source of contaminants in itself. If an incorrect type of oil is used to lubricate the compressor or if inadequate ventilation causes the compressor to overheat, it is possible that the oil may reach a sufficiently high temperature for spontaneous combustion ('flashing') to occur. This combustion produces a number of undesirable contaminants including carbon monoxide.

Even with correct maintenance the quality of the compressor output should be periodically checked using a proprietary air test kit. The BSAC Air Purity Standard recommends that, for compressed air used for diving, the level of carbon monoxide should not exceed five parts per million.

Cigarette smoke contains a significant amount of carbon monoxide. A smoker's inhalations can contain over 400 parts per million of carbon monoxide, which will take in excess of eight hours to be eliminated from the body. A pre-dive cigarette will therefore reduce the performance of the diver's circulatory system and may prejudice his capacity for exercise should a demanding situation arise. It is sound advice to abstain from smoking for several hours before diving.

Symptoms

As the effect of inhaled carbon monoxide is to reduce the blood's capacity to transport oxygen it is not surprising that the symptoms are similar to those of hypoxia. As the level of carbon monoxide increases so does the severity of the symptoms. The progression of symptoms is as follows:

Headache > Dizziness > Breathlessness upon exertion> Confusion> Nausea> Collapse > Unconsciousness > Coma > Death.

During the ascent symptoms are likely to get worse as the falling partial pressure of oxygen reduces the amount of oxygen in solution in the plasma, and hence increases the tissue hypoxia.

The most common symptom by far is the headache, and if several divers, whose cylinders have all been filled from the same compressor, complain of a headache it is a good indication that the compressor should be checked.

Carboxy-haemoglobin is bright red in colour and can give the casualty's lips and nail beds a red appearance. In practice this indication is far from obvious, particularly if the casualty is cold, and should not be relied upon.

First Aid

The immediate treatment is to stop the casualty breathing contaminated air. This will at least stop the condition from getting worse, although the slow rate at which carbon monoxide is eliminated from the body means that recovery will be equally slow.

The administration of 100 per cent oxygen will help to offset the effects of the carbon monoxide by forcing more oxygen into solution in the plasma. The casualty should be kept laid down, quiet and warm.

Severe cases should be rapidly transported to a medical facility where hyperbaric oxygen can be administered. The casualty should be continuously monitored and CPR (preferably oxygen enriched) applied should the need arise.

Mild cases should be restrained from any exertion whatsoever and advised to rest. Further diving should not be contemplated for at least twenty-four hours for mild symptoms and medical advice sought before recommencing diving after serious symptoms.

Exposure

The human body can only tolerate a small ambient temperature range. When the temperature deviates outside this range action must be taken to protect the body. When the temperature falls, insulation is required to conserve body heat, while in elevated temperatures active cooling is required. If inadequate precautions are taken, the body's metabolism attempts to compensate for conditions, which are beyond its capacity, and the body's condition deteriorates. When this happens the body is said to be suffering from exposure or hypothermia. Conversely excessively high temperatures can lead to hyperthermia.

Hypothermia

Water is an excellent conductor of heat, approximately twenty-five times faster than air. If an unprotected diver is immersed in water, which is at a temperature below approximately 33–35°C, the water will conduct body heat away faster than the body can produce it. Over a period of time, therefore, the body temperature will fall. When the body core temperature (normally approximately 37°C) has fallen to below 35°C, the diver is said to be suffering from hypothermia. The colder the water the quicker will be the fall in body temperature, and if allowed to continue unchecked the result could ultimately be fatal.

Survival time varies with water temperature, the effectiveness of the diver's protective clothing and the diver's body build. In practical terms, hypothermia is really only a potential problem in temperate or cold water climes. Even then, with adequate protective clothing it is unlikely during the course of a normal dive. Should the diver have to spend a protracted period in the water, however, or should he suffer additional cooling due to wind chill, then hypothermia becomes a real consideration.

Prevention

The simple method of prevention is to ensure that adequate protective clothing is worn at all times, not just while diving, but also before and after. A diver who feels cold before entering the water is only going to get worse while diving. Even on a calm day the motion of the boat through the water can produce sufficient breeze over the deck to cause significant wind chill, particularly on a wet diver. If shelter from the wind is not possible, a windproof jacket will greatly reduce the wind chill effect.

The body's ability to produce heat depends on its physical condition. A fit and properly fed diver will be best prepared to resist the cold, while a diver who has skipped meals to get to the diving site, and is still suffering from the 'morning after the night before', will soon become chilled.

Symptoms

Feeling cold is the first sign that a diver is losing heat faster than his body can generate it. Shivering then commences to cause an increase in bodily heat production to compensate for the heat loss. As more heat is lost the body makes a strategic withdrawal by reducing blood flow to the peripheries to conserve heat in the body core. The extremities therefore become cold to the touch, bluish in colour, and numb.

As the blood pool is increased in the core, the body tries to compensate for the extra core fluid volume by passing water. This effect is further exaggerated by immersion in cold water. Further cooling sees the initial increase in metabolic rate (the shivering) reduce and respiration rate, heart rate and blood pressure all fall as hypothermia sets in. Difficulty in concentration, apathy, confusion, and muscular stiffness are all experienced before the diver lapses into unconsciousness and convulsions. Breathing and pulse become irregular before death occurs.

Treatment

Many of the above symptoms would only be noticeable to the diver affected. Unfortunately, by the very nature of hypothermia he would be least likely to appreciate their significance. The one symptom, which is visually apparent to all, is shivering, and any diver seen to be shivering badly should be encouraged to take immediate steps to get warm again. Should this occur underwater the dive should be abandoned. Apathetic or confused responses to questions or signals are further indicators of the severity of the condition.

Once a diver is identified as suffering from hypothermia the first priority is to prevent further heat loss. He should be removed from the water and wrapped in as much insulation as possible in a position where he has the maximum protection from wind, rain and spray. In an open boat this is best achieved by laying the casualty as low in the boat as possible. A windproof and waterproof garment or a survival bag/space blanket will provide additional protection from the cooling effects of evaporation and wind chill. As a considerable amount of heat is lost from the head this should also be protected.

Once the casualty can be placed in a warm sheltered environment any wet clothing should be removed and replaced with warm, dry clothing. If hypothermia is the only problem he should be given a warm high-energy drink (e.g., glucose) to help replace lost fluids. If on the other hand he has also suffered some injury, which will possibly require hospital treatment including an anaesthetic, he should be given nothing by mouth.

On no account should the casualty be given alcohol, nor should his skin be massaged in an attempt to generate warmth.

If conscious, the casualty should be laid down with his legs raised to reduce the possibility of a reduction in blood pressure. He should be handled very gently and kept as still as possible. If unconscious but breathing, the casualty should be placed in the recovery position. For a non-breathing casualty, AV and, if necessary, CPR take priority. Some suitable insulation should be placed underneath the casualty to prevent heat loss to the surface on which he is laid.

In extreme cases of hypothermia the casualty will

need to be rewarmed slowly. This will require speedy transport of the casualty to medical treatment.

In addition to all the above protective measures, the casualty's breathing and pulse should be continuously monitored and AV/CPR instituted should they become necessary, taking precedence over all other measures.

Hyperthermia

The world's seas do not attain temperatures sufficient for the effects of heat to be a problem while actually diving. In hot climates, however, the diver may well be subject to such temperatures before or after diving. Even in temperate climates, however, protective clothing, which is comfortable underwater, can cause overheating while preparing to dive. Heat stroke and heat exhaustion are therefore possibilities, which should not be entirely discounted.

Prevention

In hot climates, long periods in direct sunlight should, if possible, be avoided. Where this is unavoidable, suitable light clothing should be worn. In particular the head and neck should be protected.

When air temperature or sunshine make the wearing of protective clothing on the surface uncomfortably hot, spend as little time as possible kitted up prior to the dive.

Plenty of fluids (non-alcoholic!) should be drunk to replace those lost by perspiration.

Symptoms

The most likely condition to be encountered by divers is heat exhaustion. This is characterized by profuse sweating, weakness, nausea and dizziness. Body temperature will be fairly normal.

Heat stroke is a life-threatening condition, which is unlikely except in very high temperatures. The casualty will have a high body temperature and his skin will be hot, red and dry, as the body's sweating mechanism will have broken down. He may become unconscious and will have a rapid pulse.

Treatment

The casualty suffering heat exhaustion should be sheltered from the source of the heat and any protective clothing removed. He should be kept quiet and made to rest. Unless the casualty is very nauseous, he should be given sips of water to drink to replenish lost fluids.

In the case of heat stroke, immediate measures should be taken to cool the body by sponging with cool water. The casualty should be laid down, kept quiet and transported as quickly as possible to medical attention.

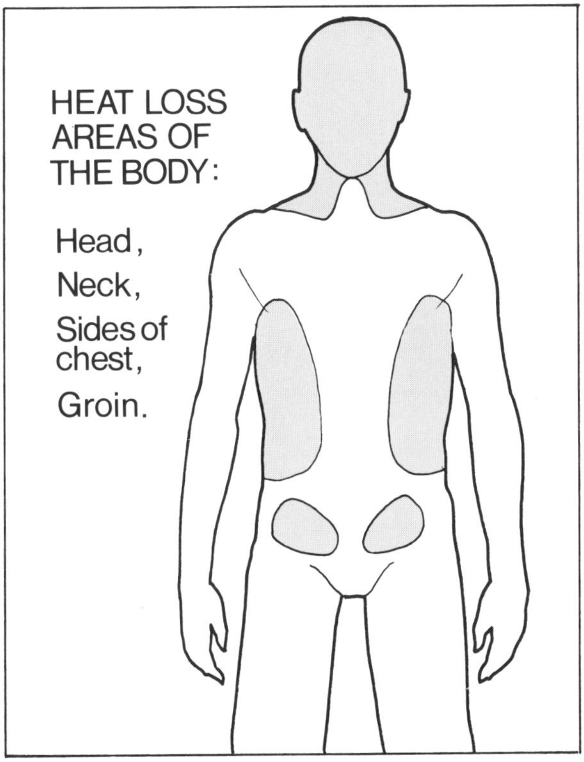

HEAT LOSS AREAS OF THE BODY:
Head,
Neck,
Sides of chest,
Groin.

Figure 121 Main areas of heat loss from body

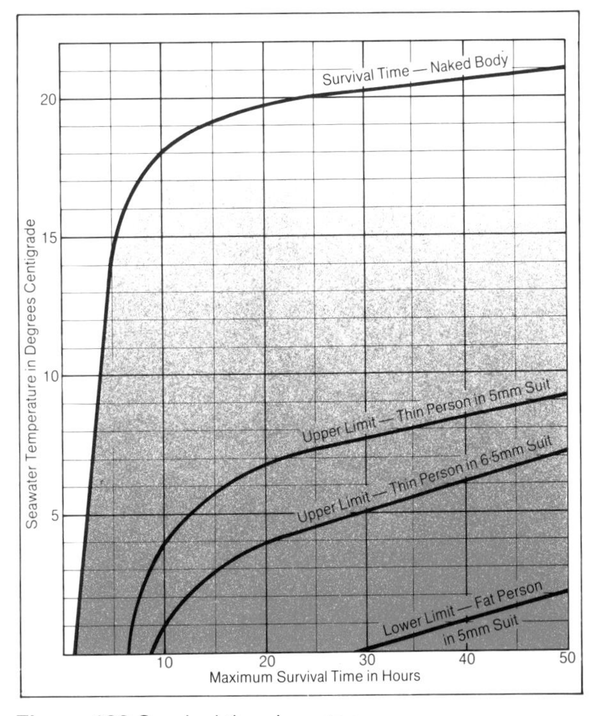

Figure 122 Survival time in water

Fatigue and Exhaustion

Energy is a result of a bodily process called metabolism. This process combines food and oxygen to produce energy and waste products. As the level of exercise an individual undertakes increases, more and more energy is required and the body's metabolism demands increasing amounts of oxygen. The individual's rate and depth of breathing increase accordingly. Depending upon the individual's physical condition, sooner or later a point is reached where his metabolism is unable to supply the required energy even at the expense of the oxygen supply to other bodily functions. The individual is then unable to respond, either physically or mentally, to further demands upon him and is said to be exhausted. The gradual deterioration of the individual's capabilities approaching this point is called fatigue.

On land, exhaustion is easily remedied by rest, but when diving any inability to respond to the demands of the situation prejudices the diver's safety and could result in drowning.

The most common cause of exhaustion is over-exertion, but in the diving environment this is further compounded by a number of other factors. Water is a viscous medium and consequently any actions require considerably more effort than the same actions performed in air. At depth, the increased pressure (and hence density) of the diver's breathing air makes it more viscous and the diver experiences increased breathing resistance. Thus his breathing effort must increase.

Prevention

The most obvious and effective way to avoid exhaustion is to minimize the exertion required while diving. Each diver should know his personal limits and keep within them. Where heavy exercise is unavoidable, such as when carrying bulky equipment, swimming against a tidal stream or when performing a rescue, it is important that the diver paces the effort required to remain within his capabilities.

All breathing equipment should be regularly maintained by a competent facility to ensure that the airflow resistances are correct. All other personal equipment should fit comfortably and without tightness or restriction, and any protective clothing worn must be adequate for the diving conditions.

The effort required in kitting up prior to a dive should not be underestimated, particularly where large heavy cylinders are concerned. Kit up in good time for the dive such that all equipment can be put on at a relaxed pace without leaving you standing fully kitted for an extended period before entering the water. Kitting up in haste will not only prevent you entering the water in a relaxed state but vital equipment or checks may be forgotten.

Overweighting and inadequate dive planning are probably the two most common reasons for incidents where diver exhaustion is the root cause. Proper weighting and conscientious buoyancy control during changes of depth will ensure that, throughout the dive, energy is not wasted. Laziness in adjusting the amount of weight on the weightbelt, and relying on compensating by excessive buoyancy adjustment causes a number of incidents each year. The dive itself should be well planned to ensure that the divers are not subjected to water conditions of current or swell which are beyond their capabilities.

Figure 123

Symptoms

As fatigue sets in the diver's breathing will increase in rate and depth and he will begin to feel an increasing weakness or tiredness. If at this stage nothing is done to alleviate the situation, the diver's breathing will become laboured and further deterioration will cause confusion and anxiety. Taken to extremes, the diver is likely to take irrational actions or to panic.

The initial symptoms are, fortunately, noticeable not only by the casualty, but also by his buddy, who can detect the increased respiration by the casualty's exhaust bubbles. Corrective action should be taken at this stage and the casualty's condition should not be allowed to deteriorate further.

Corrective Action and First Aid

As soon as the initial symptoms of increased respiration are noticed it is important that all physical activity ceases and the diver is allowed to rest and regain his composure. Underwater the diver should settle on the bottom and hang on to some solid object to counteract any tidal stream current. If composure is quickly regained the dive may then continue but at a reduced pace to prevent reoccurrence of the problem.

If composure either cannot be regained or takes an excessive time, then the dive should be abandoned and the surface regained. A controlled buoyant ascent will allow the casualty to surface with minimum effort, but the casualty's buddy should closely monitor him and be prepared to assist if necessary.

On the surface the casualty should inflate his BC to provide support and to allow him further rest. If the sea is very choppy the casualty should remain breathing from his demand valve to prevent the inadvertent swallowing of sea water and consequent choking.

If boat cover is available, the divers should signal to be picked up. Once alongside the boat the casualty's buddy should help him to remove heavy equipment before he attempts to climb into the boat. Once the heavy equipment has been removed, deflating the BC will ease entry into the boat. Occupants of the boat should help the casualty inboard.

Should a swim to an anchored boat or to shore be unavoidable, the casualty's buddy should do as much as possible to assist him and be prepared to tow him if necessary. If anything other than a short swim on a calm surface is required, serious consideration should be given to jettisoning heavy equipment. Exit from the water through surf will be particularly difficult for the exhausted casualty, and if possible assistance from people on shore should be summoned to get the casualty clear of the water quickly.

Once safely out of the water, the casualty should be made to rest and be kept warm. He should be monitored until he has clearly regained his composure, as he is likely to be suffering from mild shock.

Should a diver complain of weakness or exhaustion after surfacing, it should be remembered that this is also a common symptom of inadequate decompression. The dive history should therefore be checked before assuming that exertion is the only cause.

Figure 124 Removing a tired divers mask to reduce stress

Figure 125 Tired diver being towed by buddy

Aftercare

Once a casualty has been removed from the water further attention will be necessary, the nature and extent of which will depend upon the condition of the casualty. This attention will need to continue until such time as the casualty is fully recovered, if his condition is of a minor nature, or until he can be delivered into the care of appropriate medical personnel for more serious conditions. In parallel with these activities other actions will need to be taken to contact the emergency services and otherwise manage the situation, as covered in the following sections of this manual. (see also Page 49).

The Unconscious, Non-breathing Casualty

The requirement to administer AV dictates that the casualty must be laid flat on his back. Should Chest Compressions also be required this must be on a hard surface. When padding is used under the casualty's shoulders to assist in achieving a good neck extension, this should be the minimum necessary to obtain a clear airway.

When moving or positioning the casualty particular care should be taken to protect the head.

The Unconscious, Breathing Casualty – The Recovery Position

The recovery position is a stable three-quarters prone position suitable for the placing of an unconscious but breathing casualty. The placing of the arms and legs ensures that the position remains stable even on a moving surface, such as in a small boat. The head is placed in such an attitude that good neck extension is obtained to provide a clear airway and is turned just sufficiently to keep the face clear of the ground. This provides a drainage path from the mouth to clear any vomit, while keeping the mouth and nose clear of the surface to prevent inhalation of the vomit.

The Conscious Casualty

Except for minor conditions, the conscious casualty should be laid down and kept quiet. Conscious casualties who may have difficulty in maintaining their airway, such as those suffering from extreme exhaustion or exposure should be laid in the recovery position.

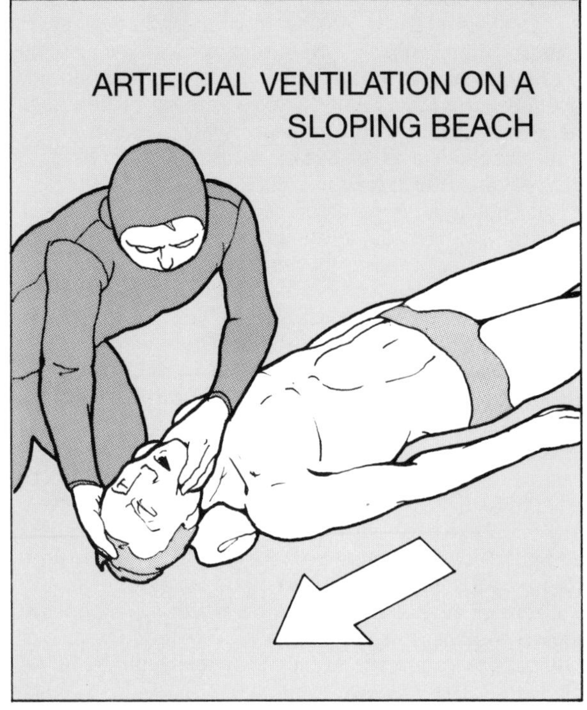

Figure 126 Using a sloping beach for artificial ventilation

Figure 127 Casualty in recovery position

General Considerations
All casualties will, to a greater or lesser extent, be suffering from shock, which should receive the appropriate attention, in addition to any other treatment they may require. Shock and its treatment are fully described elsewhere in this manual.

As the casualty will most likely still be wet, appropriate measures should be taken to protect him from the cooling effects of wind chill and evaporation. If he cannot be removed to a sheltered environment, the casualty should be covered with a windproof garment or placed in a survival bag or space blanket.

An evaluation of the casualty's overall condition should be made at the earliest opportunity after he has been removed from the water or, where symptoms are delayed until some time after leaving the water, as soon as any symptoms occur. Such an assessment is described in the First Aid section of this manual. The casualty should be constantly monitored, and any changes in his condition noted until he can be delivered into the care of medical personnel. Any information regarding changes in the casualty's condition during this period may assist medical personnel during the subsequent treatment of the condition. Should the casualty's pulse or respiration fail, AV and, if necessary, chest compressions must be commenced immediately.

Figure 128 Administering oxygen on a beach

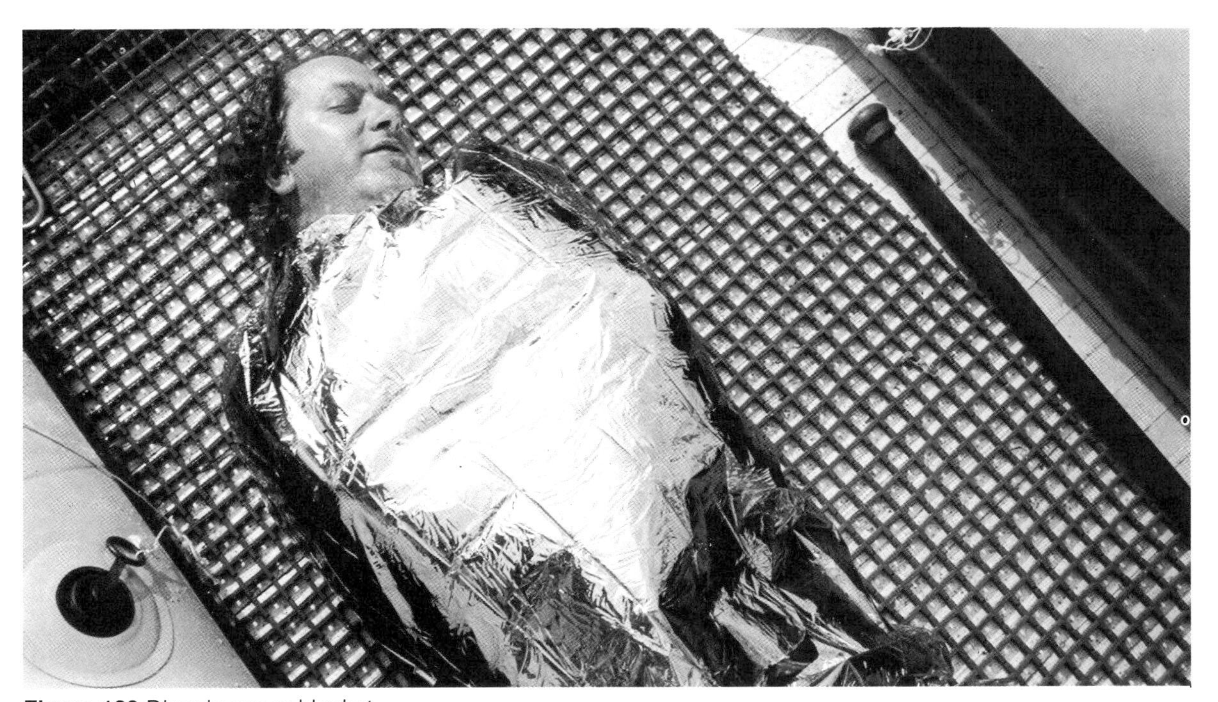

Figure 129 Diver in space blanket

Emergency Services

The requirements for medical help following any accident mean that there will be a need for the use of the available emergency services. These services may involve several organizations with different functions. This section is aimed at acquainting the casual user with the facilities which may be needed.

HM Coastguard

HM Coastguard is responsible for initiating and co-ordinating all civil maritime search and rescue around the United Kingdom and part of the Eastern Atlantic (between latitudes 45 degrees and 61 degrees North, and as far as Longitude 30 degrees West). The area is divided into six Maritime Search and Rescue Regions (MSRRs), each supervised by Maritime Search and Rescue Co-ordination Centres (MRCCs) based at Aberdeen, Great Yarmouth, Dover, Falmouth, Swansea and the Clyde. The 'Shannon' area is also included to cover the Republic of Ireland. Each region is divided into districts with a Maritime Rescue Sub-Centre (MRSC) in each. Auxiliary Coastguards assist the Regular Coastguard Officers with the manning of watch and rescue stations around the coast. All MRCCs and MRSCs maintain a radio watch on VHF Ch 16 and 2182 kHz, and are linked to telephone and telex networks. Some stations also have a radar watch facility, including the detailed monitoring of the Dover Strait. Some stations can also offer a VHF position fixing capability by the use of direction-finding equipment in adjacent stations. A visual search is maintained in times of bad weather at some 100 Auxiliary Watch Stations. The Coastguard also has a cliff and beach rescue role with fully equipped motor vehicles. HM Coastguard acts as the co-ordinating and controlling centre in any rescue or medical evacuation following an accident, and can be consulted by telephone (999 and ask for the Coastguard) or by VHF radio (Ch 16). The Coastguard will arrange for medical advice on decompression incidents or other accidents, and will liaise with the other services for subsequent actions.

Royal National Lifeboat Institution

The RNLI is a charitable organization supported entirely by voluntary contributions. There are about 200 RNLI stations around the coast of the United Kingdom, the Republic of Ireland, the Isle of Man and the Channel Islands. About 130 lifeboats and a similar number of inflatables are based at these stations, although some of the inflatable boats only have a summer and daylight operating capability. They are manned by volunteer crews; usually each station has one full-time RNLI employee who may be the motor-mechanic or coxswain-mechanic. Offshore lifeboats maintain a watch on 2182 kHz and VHF Ch 16 when on service; the inflatable boats use VHF Ch 16. They can also use other channels for communication with other services. Lifeboats now carry a quick-flashing blue light.

Figure 130 An H.M. Coastguard Station

Figure 131 An RNLI Station

Helicopter Rescue

Air-Sea rescue helicopters can be requested by the Coastguard to support a search or casualty evacuation. These helicopters are provided by the Royal Navy, the Royal Air Force or Coastguard. The helicopters will be controlled by a Rescue Co-ordination Centre at Edinburgh or Plymouth, and have a role to cover Service as well as civilian distress. The RAF can also call upon fixed-wing aircraft for longer or more extensive searches. The endurance of the helicopters is limited, as are the conditions in which they can operate, only some having a night-flying or bad-weather capability. Evacuation of a casualty from a boat may require the helicopter to operate within an altitude of 8m (20ft), although higher operation is possible for some aircraft. For information on helicopter evacuation (see Page 100).

Figure 132 A Rescue Helicopter used by the RAF

Ambulance

The Area Health Authority is responsible for the provision of the ambulance service for the transport of sick or injured people. The ambulance service can be contacted by telephone (999 and ask for ambulance) and will normally respond in less than thirty minutes. The actual time for response depends upon the location of the casualty, as well as the availability of equipment. Many ambulances are equipped with oxygen equipment compatible with the BSAC recommended equipment. However, it is advisable to specify that oxygen is required for a diving or drowning accident. All ambulance personnel are trained to an advanced level of first aid and some, in addition, are trained in advanced techniques beyond the normal province of first aid. Some ambulances are also equipped with specialized equipment such as cardiac monitors with associated relay equipment. This equipment could be of great value during post-accident treatment of diving incidents. Ambulances are normally instructed to take the casualty to the nearest available emergency unit. This may cause some difficulty in the case of diving accidents, since a normal emergency unit is not equipped to deal with recompression. Transport to other facilities may require authorization from a doctor or the Coastguard following specialized consultation.

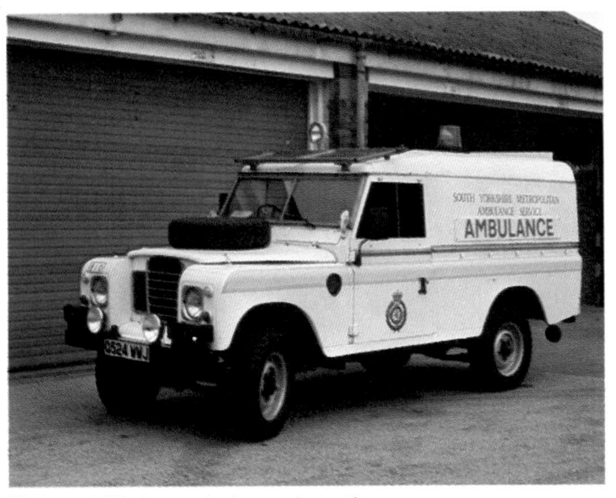

Figure 133 An ambulance in action

Police

The Police will not normally be involved in rescue situations unless there is a major evacuation of casualties to be arranged. They will, however, be able to assist in the clearing of access routes for emergency vehicles and will provide an escort if needed for vehicles transporting casualties. Some Police forces are equipped with a helicopter division that may be involved in the evacuation of casualties from a shore-based location, although they are not normally equipped to the same level as the Services helicopters for Air-Sea rescue duties. The Police will also be involved in any fatality, seeking statements from all parties involved in the accident.

Transportation

Prior to arranging transportation the primary consideration must be the well-being of the casualty, to ensure that there is no deterioration in his condition. The techniques of moving an injured person are best left to trained personnel unless there is immediate danger to life, or if access to skilled help is not available. This section will deal with the requirements of moving an injured person, although the injuries may vary and no specific examples can be quoted. The first consideration, once the decision to move the casualty has been made, must be the manner of movement. This will depend upon the facilities and assistance available, and the nature of the injuries to the casualty. No attempt should be made to move a seriously injured casualty if insufficient help is available. Before the casualty is moved, a quick but thorough examination must be performed to ensure that the nature of the injuries is fully understood. The details of this examination should be retained, preferably in a written form, so that the information can be passed on.

General Principles

Before moving the casualty, he should be informed of exactly what is going to happen, as this will reduce apprehension. All those involved in the move should also be thoroughly briefed on the sequence of moving and their relative actions.

Casualties with serious injuries, those that have required any form of resuscitation, barotrauma, or decompression illness should be transported by ambulance or similarly equipped transport. Only those suffering from minor injuries, or involving only injuries to the upper limbs, should be taken by car. Whatever method of transport is to be used, the aim is the same, to transport the casualty to the destination without deterioration or discomfort. In most cases the casualty should be transported in the same position as has been used up to this point, this will normally be the most comfortable position for the casualty. The condition of the casualty should be monitored throughout the move and, if necessary, the journey halted or modified. It is more important to stabilize the casualty's condition than to speed towards treatment.

Figure 134 Neil Robertson stretcher used by helicopters

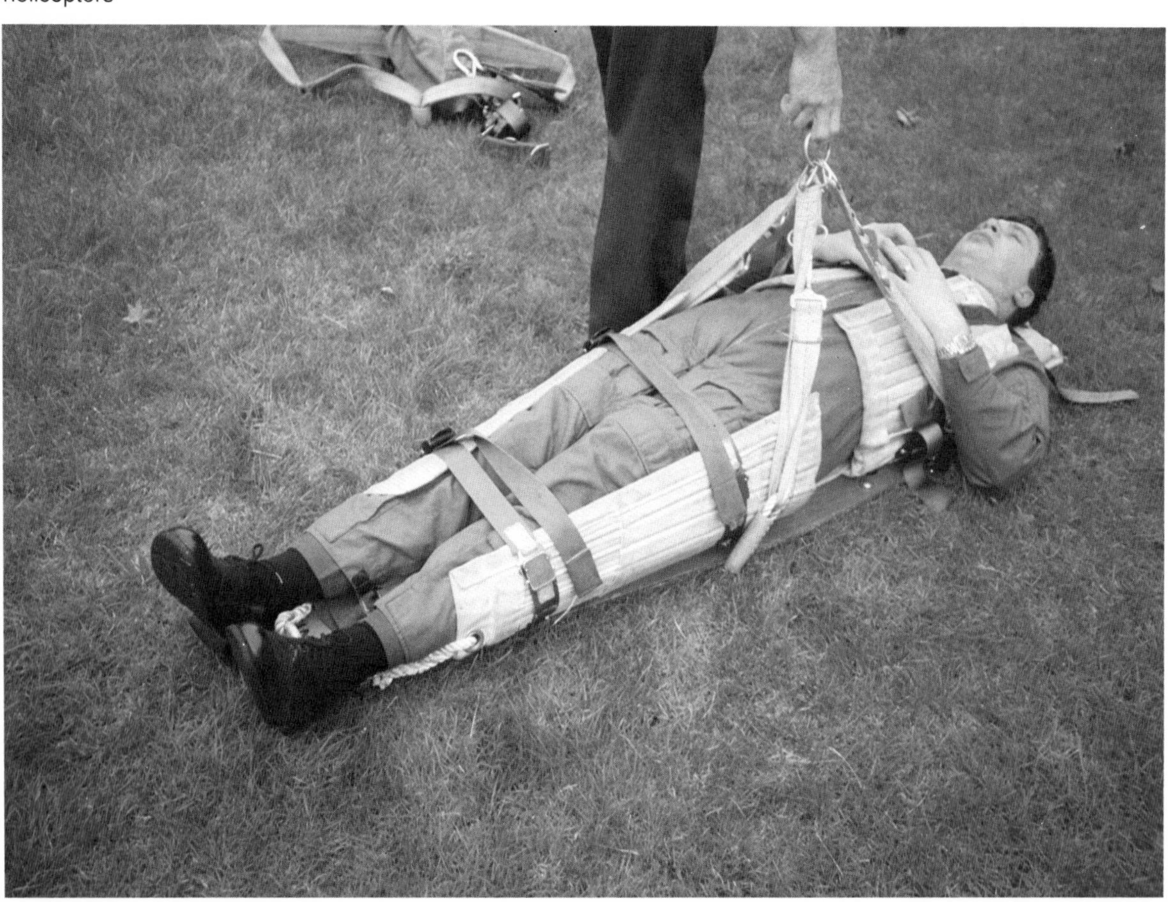

Carrying a Casualty

If it is necessary to move a casualty from one area to another, it is important to think carefully about the method to be used. The choice of method depends upon the assistance available, the nature of the injuries, and the relative size of the casualty. Simple methods of moving a walking, although injured, person include using the rescuer as a human crutch, standing on the injured side of the casualty and placing the casualty's nearest arm around the rescuer's neck (see Figure 102). This method cannot be used if there are injuries to the upper limbs. The Fireman's lift can also be used to move a conscious or unconscious child or slimly built adult, but care must be taken to ensure that no internal injuries are aggravated, and that the rescuer can cope with the weight of the casualty (see figures 103 and 105). If more than one person is available to transport the casualty it may be possible for one person to carry the torso and one the limbs, in a manner similar to that for the drag method of removing someone from the water.

Stretchers may also be used if available. If supplied by the emergency services full instructions will be given by the crew concerned, and it is important that these are followed carefully. The most common form of stretcher may be a modified Neil-Robertson stretcher, or a cradle carried by helicopters and cliff-rescue teams. These stretchers require careful utilization to ensure the safety of the occupant (see Figure 134). Ambulances will usually use a trolley bed in which the casualty is strapped by the crew. These trolleys are designed to lock into an ambulance and require careful use by untrained personnel.

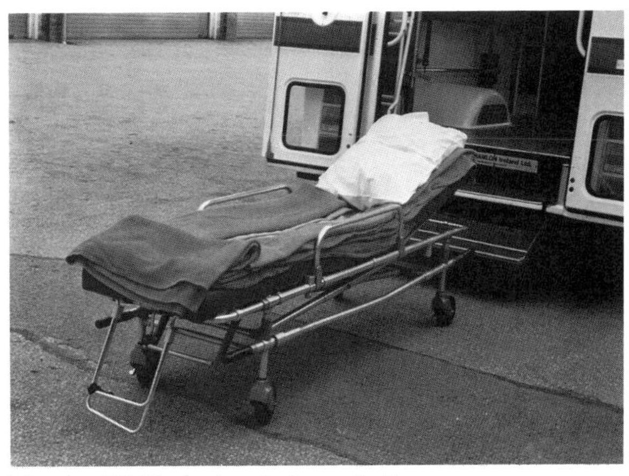

Figure 136 Typical ambulance trolley

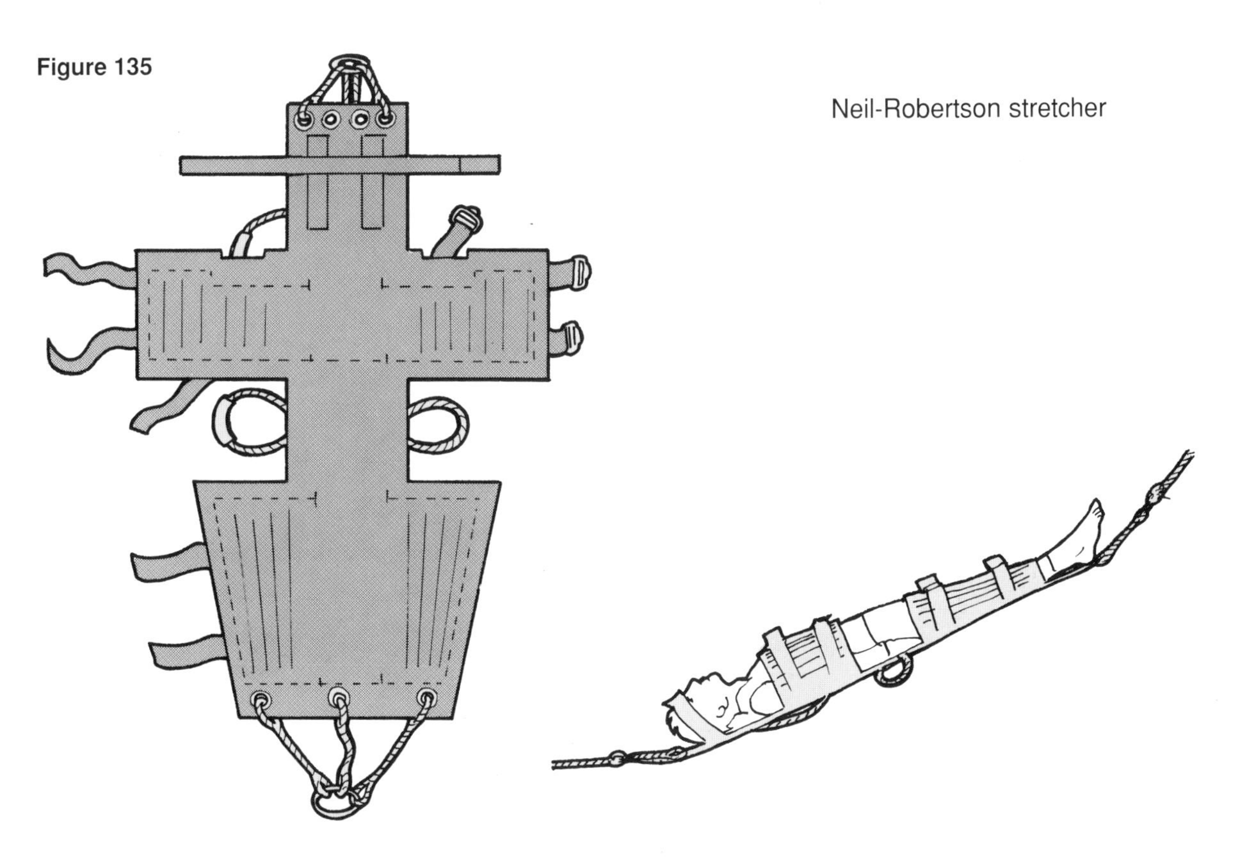

Figure 135

Neil-Robertson stretcher

Evacuation by Helicopter

This may be the most frequent method of evacuation for a casualty from a boat offshore, but it is also one that may require careful thought and even some modification of the boat's structure. Due to the limited lifting capacity of helicopters they may have to hover within 8m of a boat containing a casualty. The considerable downwash may also cause problems to the craft concerned. While hovering, the pilot has no vision of the craft below him and must rely on verbal instructions from his winch operator. The procedure for evacuation must therefore include assisting the crew as much as possible without causing hindrance.

The first priority must be to assist the approaching pilot to identify the craft by use of suitable flares or daylight smoke. VHF radio can also be used via the Coastguard or the nearest coastal radio station to contact the aircraft. Although the helicopter may carry the marine band VHF radio, the crew normally do not monitor routinely. If the boat has an elaborate superstructure it may be necessary for a casualty to be lifted from an inflatable dinghy streamed behind the main craft. As the winchman approaches it is advisable to keep well clear, as helicopters develop static electricity during flight and this needs to be earthed before contact can safely be made. This is normally done by dipping the winch wire prior to contact. This wire must not be secured to the boat in any way.

The normal way to lift a casualty is by use of a strop put over the head and shoulders and fitted under the arms, with the padded part in the small of the back. The sliding sleeve is used to tighten the strop (see Figure 137). Once this has been done, the signal to raise is passed to the winchman using the standard thumbs-up signal. If the winchman descends with the strop his instructions should be followed implicitly, especially if this involves the Neil-Robertson stretcher. In some cases the helicopter pilot may opt to use a Hi-Line technique, where a weighted line is lowered attached to the winch wire. This line (usually 50m in length) is used to guide the main winch wire to the craft. It contains a weak link that will snap if undue pressure is exerted on the Hi-Line extension. It is therefore important not to pull or tie the line onto the craft, but merely to use it as a guide. This guide is reclaimed by the helicopter when the casualty is winched aboard. The casualty will be raised until level with the door, and then recovered inside. Instructions from the crew must be followed carefully.

It is important to remember that the costs of operating a helicopter for Air-Sea rescue duties are extremely high, so time must be utilized to the optimum. Evacuation by helicopter may also incur an increased risk due to the lower ambient pressure at altitude for casualties suffering from decompression illness. It is therefore important for the maximum altitude to be limited normally to 300m (1000ft), which may cause problems in navigation for the crew and lead to extensive detours being required.

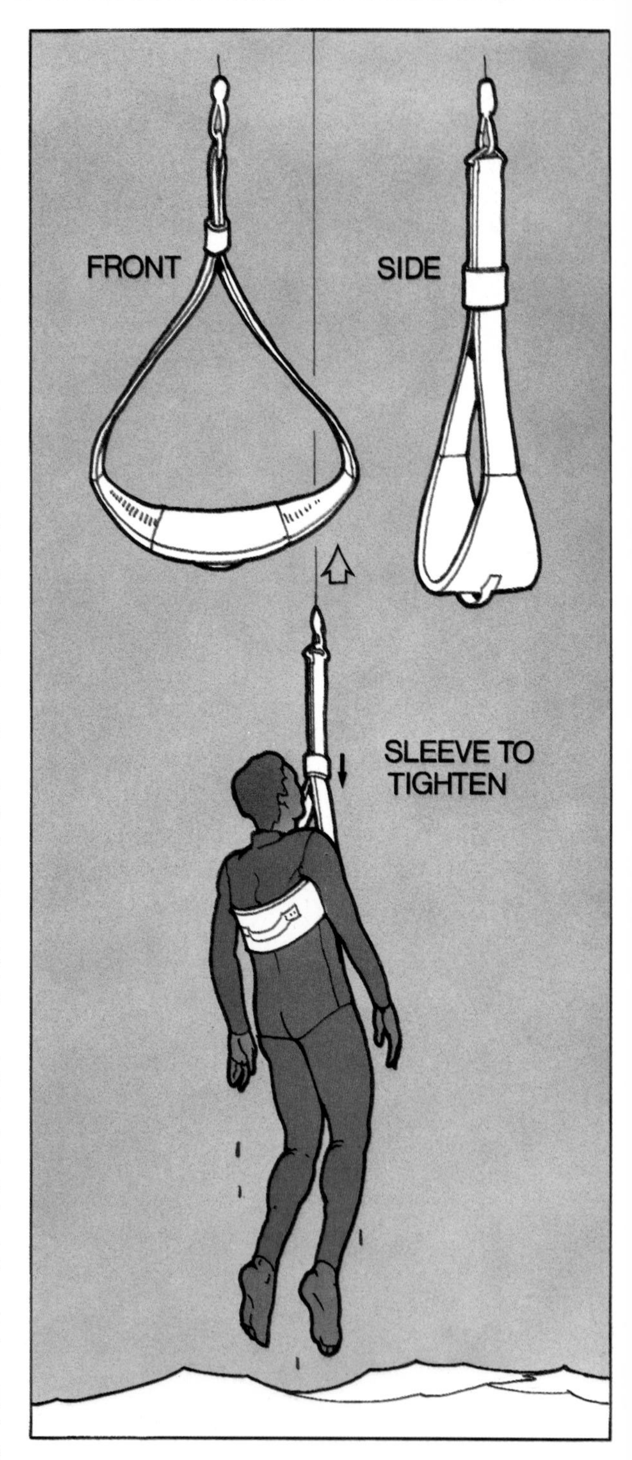

Figure 137 Recovery sling used by helicopters

Figure 138a A rescue helicopter in action

Figure 138b Diver being winched aboard helicopter

Evacuation by Lifeboat

If the decision is made to evacuate the casualty by lifeboat, the vessel should be put into a position that will allow the lifeboat to identify and then come alongside in order to take off the casualty. VHF radio can be used to communicate with a lifeboat, although contact is normally established via the Coastguard service The crew's instructions must be followed, and full information given to them to assist the treatment of the casualty's injuries.

Figure 139 An RNLI Lifeboat

Figure 140

Rescue Equipment

Oxygen Equipment

Having established that the administration of oxygen can be beneficial in certain diving emergency situations we must then consider which type of equipment will best meet our needs. We must first point out that the type of equipment most often found in ambulances and first aid posts is unsuitable for the majority of diving accidents. We therefore require equipment that is specifically configured for diving needs, the most important feature being that it must be capable of supplying the casualty with 100% oxygen for a period of time.

The idea of adapting existing diving equipment for this purpose is not advisable since the resulting arrangement would fall short of the requirements in a number of ways.

The complete set of equipment which we require will consist of the following main components:

1. A cylinder.
2. A pressure reducer.
3. A connecting hose.
4. One or two demand valves.
5. A mask for each demand valve.
6. A water-resistant carrying case

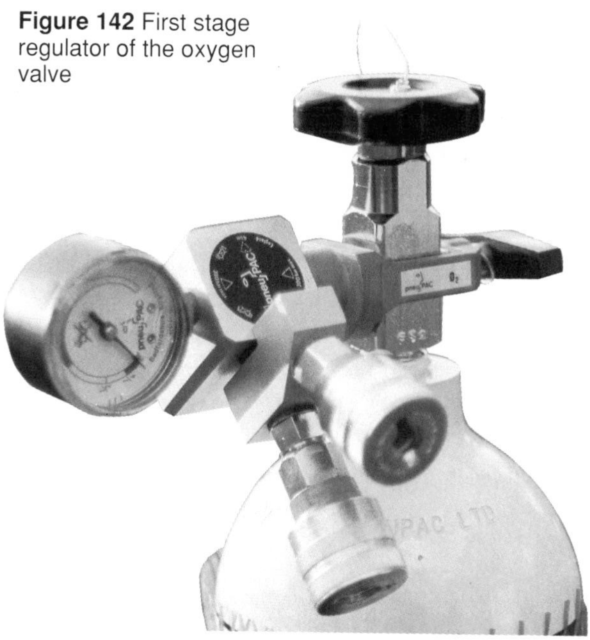

Figure 142 First stage regulator of the oxygen valve

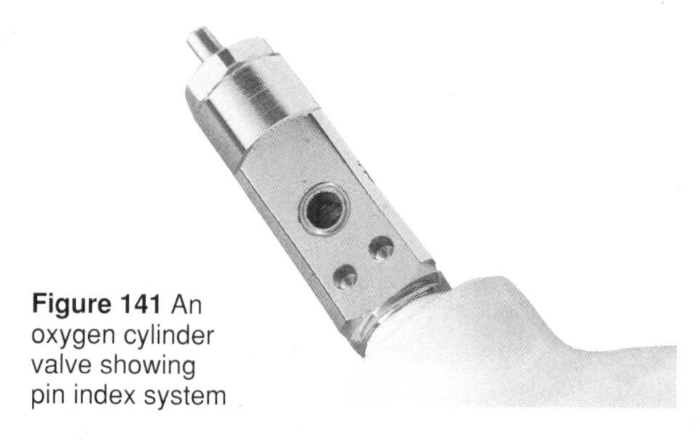

Figure 141 An oxygen cylinder valve showing pin index system

The Cylinder

The supply of oxygen under pressure must be both portable, and sufficient to support the treatment. In practice a 'D'-size cylinder is found to be most suitable since it has a capacity of 360 to 415 litres, depending on whether it is made from aluminium alloy or steel, and is compact enough to be carried on small boats. Larger boats and permanent sites will have the possibility of a larger store of oxygen, but consideration should be given for the possible need for the equipment to accompany a casualty under treatment while being transported to a hospital or recompression chamber. Further 'D'-size cylinders will provide the best means of continuing or extending treatment.

The cylinder should be kept with the oxygen equipment and not stored with air cylinders. It is identifiable by a different colour coding from that used for air, black body with white shoulder in the UK, green overall in the USA.

The portable sizes are fitted with a pillar valve with a pin-index connection system which avoids the possibility of confusion with air systems. It will not accept a standard air regulator.

Pressure Reducer

This is equivalent to the first stage of a diving regulator in that it reduces the pressure of the oxygen supply to that required by the demand valve or constant-flow system. It has a yoke which matches the pin-indexed connection of the cylinder valve. The seal is effected by a flat washer which will need to be checked periodically to ensure that it is (a) in place and (b) providing an effective seal. It is important to keep spare washers as a damaged washer could render the entire equipment inoperative.

The reducer should ideally be fitted with two outlets, each with quick-connect systems. These outlets will supply either two demand valves, or one demand valve and one constant-flow mask. In some systems the constant-flow outlet will be dedicated to this purpose since it will contain the flow-limiting mechanism, while in other systems the flow limiter will be incorporated in the hose connector. The constant-flow system, when in use, must provide at least 10 litres per minute of oxygen.

The reducer should also be fitted with a pressure gauge which gives a constant indication of the pressure remaining in the cylinder.

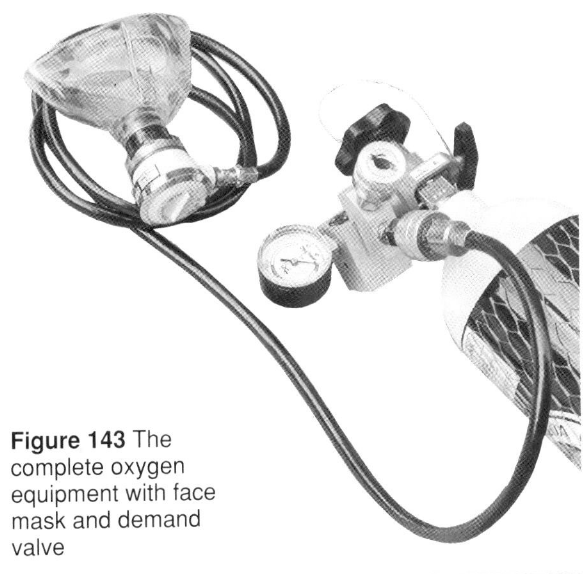

Figure 143 The complete oxygen equipment with face mask and demand valve

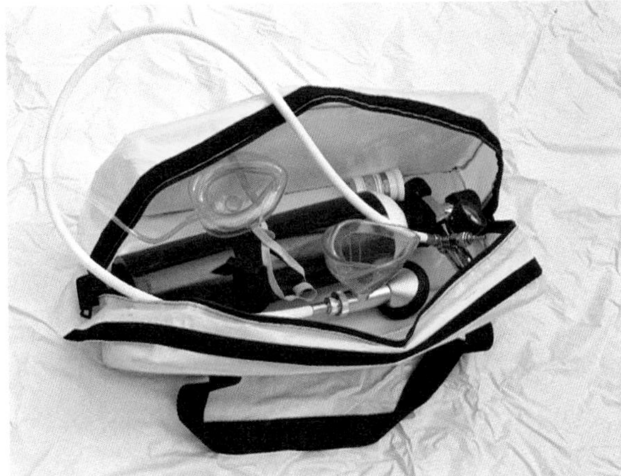

Figure 144 Portable oxygen therapy equipment

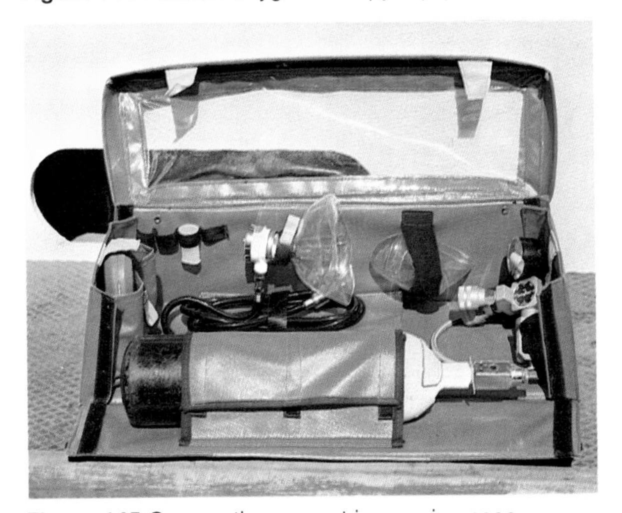

Figure 145 Oxygen therapy set in carrying case

Figure 146 Administering oxygen to two casualties (above) and to one casualty (below)

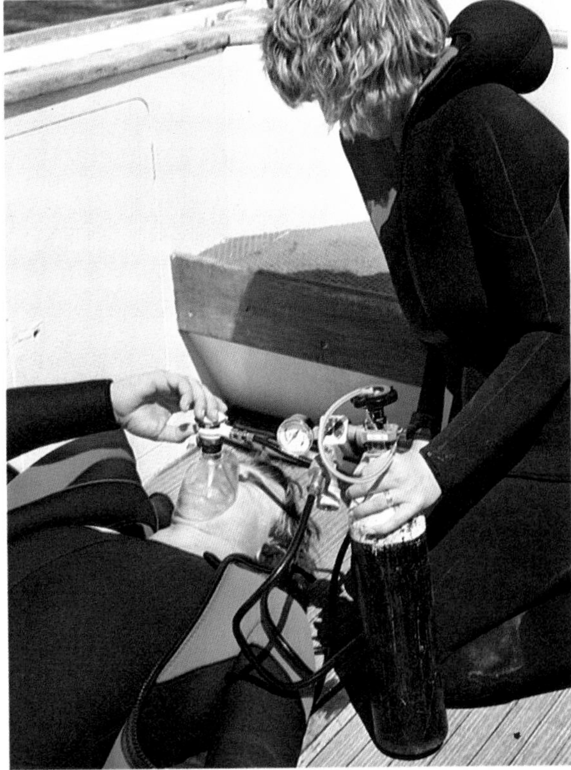

Connecting Hose
The hose which connects the reducer to a demand valve will be a medium-pressure hose, usually directly connected to the demand valve, with a quick connector at the other end. The hose used to connect a constant flow mask to the reducer will be a much lighter and thinner flexible tube. The two types of hose are not interchangeable and should not be confused.

Demand Valve
As with a diving regulator, the demand valve allows the casualty to breathe oxygen as required. It is operated by the casualty's inhalation and closes after each breath. The demand valve is essential to any treatment which requires 100% oxygen since it does not allow the flow of gas to be diluted with atmospheric air. It also prevents any wastage of gas. Some regulators are fitted with a device which allows the rescuer to inflate the casualty's lungs by positive pressure from the demand valve. This type of demand valve is not recommended for use by non-medical personnel due to the potential risk of over-inflation. When necessary the positive pressure is provided by the rescuer's own lungs through oxygen-enriched artificial ventilation.

The Mask
Fitted to the demand valve is an oro-nasal mask which will cover the casualty's mouth and nose. It is fitted with an inlet which corresponds to the demand-valve outlet, and will have a flexible seal. Masks are generally transparent to permit the casualty's airway to be monitored by the rescuer. They also come in several sizes to cope with the variations expected in casualties of different age and size.

A mask intended for use with a constant flow of oxygen should have a large orifice suitable for a rescuer to give artificial ventilation as well as a connection for the connecting hose bringing the oxygen supply. The most popular type of mask for this type of treatment is the pocket mask, which adapts well to most facial contours.

Guidelines for the use of oxygen administration equipment

1) Only administer normobaric oxygen as a supplement to normal first aid procedures to divers who are suffering a diving-related problem.
2) Keep oxygen equipment clean, dry and in a safe location.
3) Never permit oil, grease or other readily combustible substances to come into contact with oxygen equipment.
4) Use oxygen equipment only in a well-ventilated area clear of any sources of ignition (lighted cigarettes, electrical equipment, internal combustion engines, etc.).
5) Use only soap and water for cleaning oxygen equipment.
6) After use, disinfect oro-nasal masks and associated valves using disinfectant alcoholic wipes and allow to dry naturally. From time to time, give them a longer soak in a mild disinfectant solution (e.g., 'Milton' solution).
7) Check the contents and functioning of oxygen equipment at regular monthly intervals.
8) Never store oxygen equipment with either the cylinder turned on or the regulator pressurized.
9) Store oxygen cylinders in a location free from fire risk and away from sources of heat and ignition and from combustible materials.
10) Use only equipment purpose-built for the job by specialist manufacturers. Many widely used engineering materials are not safe when used with oxygen, therefore do not use home-made apparatus or make adaptations or running repairs to specialist apparatus.

Figure 147 Monitoring the casualty whilst administering oxygen

Carrying Case
If the equipment needs to be transported with the diving party, onto beaches and small boats, then it is important that the carrying case should protect its contents from the corrosive effects of seawater and should keep the equipment clean. The components should be kept secure, but must come to hand easily when required. The case will need to hold all the items listed previously, as well as a brief reminder on the methods of use, and the actions to be followed. Avoid making the equipment so heavy and unwieldy that it is too big to stow in a small boat and more likely to be left behind. Oxygen equipment is most effective when it is available at the point of surfacing of the diving party.

Oxygen equipment is very specialized and is constructed of materials which may not resist salt-water corrosion. It is designed to function with a minimum of lubricants, and its servicing requirements are therefore quite different to that for diving equipment. The sets described above are not suitable in any way for diving and should not be taken into the water.

Figure 148 Administering by constant flow

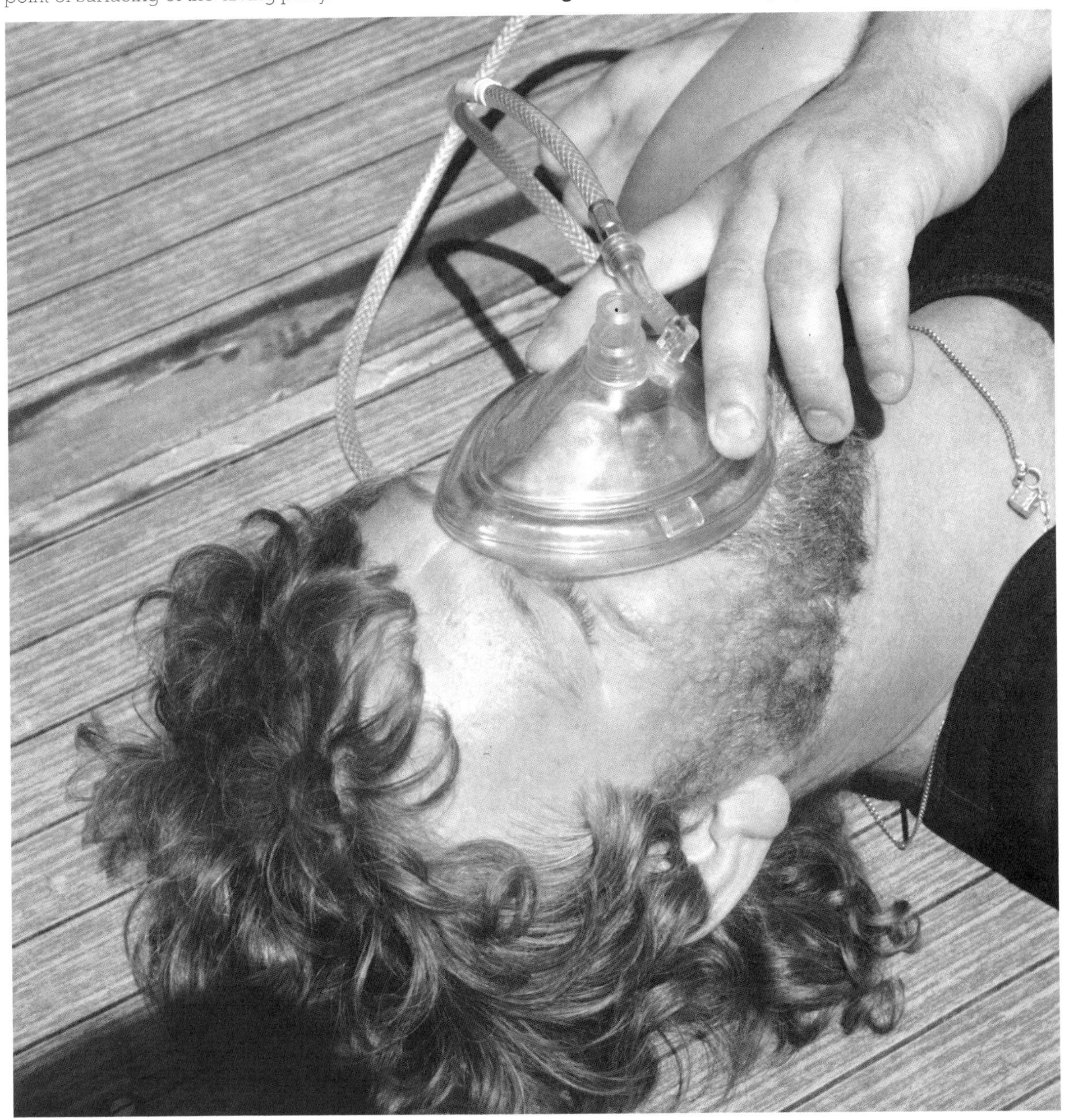

Using Oxygen

Of the casualties who will benefit from the administration of oxygen, historical incident data show that by far the majority will be breathing and, probably, conscious. A much smaller number will not be breathing. Two fundamental techniques allow oxygen to be administered to both breathing and non-breathing casualties.

The Breathing Casualty

The following technique assumes the most likely case of a casualty suffering from either decompression illness or burst lung. The most suitable method of administering oxygen to these casualties is via a demand mask. This is particularly true of decompression illness where the establishment of the most favourable pressure gradient for the elimination of nitrogen from the casualty dictates the need for 100 per cent inspired oxygen.

The casualty should be laid down, made as comfortable as possible and the administration of oxygen commenced at the earliest opportunity. The casualty will be suffering from shock as a consequence of his condition, and should be treated appropriately.

If the casualty is conscious explain to him what to do with the oro-nasal mask, and allow him to hold the mask to his face. If the casualty is unconscious the rescuer will need to hold the mask. Do not fix the mask in place on the casualty's face by any strap arrangement as the casualty may vomit, and the mask will need to be cleared from his face very rapidly. As nausea is a symptom of decompression illness this eventuality is a very real consideration and constant monitoring of the casualty's breathing is essential.

Shortly after the commencement of oxygen administration there may be a transient worsening of symptoms. This is due to the increased partial pressure of oxygen in the blood initially causing oxygen to diffuse into any bubbles more rapidly than the reduction in blood nitrogen partial pressure causes nitrogen to diffuse out. Additionally, the body's natural reaction to the increased partial pressure of oxygen results in a reduction in blood flow due to vasoconstriction.

The earlier that oxygen can be administered the more benefit it will be. If the supply is limited, rationing it in an attempt to make it last longer will only reduce its effectiveness. Similarly, if a second casualty also needs oxygen do not interrupt the supply to the first. A second mask should be used and the consequent more rapid depletion of the oxygen supply accepted. If a second demand mask is not available, the second casualty should be provided with oxygen via a constant flow mask.

Many casualties suffering from decompression illness are found to be significantly dehydrated. To counter this the casualty should be given drinks of fluid, in small quantities, at an average rate of approximately 1 litre/hour. Isotonic drinks are best but otherwise use water/squash drinks. Do not administer fluids if the casualty is less than one hour from a recompression chamber or if they are likely to vomit/inhale the fluid or if the casualty is suffering an internal injury which may require the subsequent administration of a general anaesthetic.

The administration of oxygen remains the highest priority.

It is possible that, during the period of oxygen administration, the casualty's condition will improve to the extent that he feels he has been fully 'cured'. Irrespective of this apparent improvement the casualty must still be delivered to proper medical treatment. Should oxygen administration be terminated, symptoms may subsequently reappear or other consequences of the condition, not detectable by either casualty or rescuer, may still require medical attention.

Oxygen can also benefit casualties suffering from carbon-monoxide poisoning or shock. The technique for these cases is broadly similar to that described in the preceding paragraphs, with the obvious difference that the type of medical treatment subsequently required would be different.

Figure 149 Oxygen enriched AV

The Non-breathing Casualty

Artificial ventilation must be started at the earliest opportunity. It should be initiated and continued until oxygen equipment is available. Time is so vital that provision of oxygen equipment should not be allowed to delay the commencement of AV for even a few seconds. Once the equipment is available then a transition can be made to oxygen-enriched AV by the use of a pocket mask connected to a constant flow oxygen supply.

The rescuer takes up a position above the casualty's head and the pocket mask is placed over the casualty's mouth and nose. The rescuer holds the mask in place by using both hands, one on each side of the mask. The thumb and forefinger of each hand are placed on the skirt of the mask while the remaining fingers are hooked under the jawbone. By tensioning the fingers the mask is held securely against the casualty's face. It is essential that the fingers are on the jawbone itself and not the fleshy underside of the chin, as this will interfere with the casualty's airway. By pulling the casualty's head towards him, the rescuer provides the proper extension of the casualty's neck.

Oxygen-enriched AV is then simply performed by the rescuer blowing into the central hole on the mask while the oxygen supply is connected via the appropriate elbow connection. Once the casualty's lungs are inflated, the rescuer removes his mouth and the casualty's exhalations pass out through the same central hole.

With a 10 litre/min. constant flow supply, the percentage of oxygen in the air delivered to the casualty is increased from approximately 16 per cent, as in the rescuer's normal exhalations, to approximately 40 per cent, the exact value depending upon the precise rate and depth of the ventilation.

From the rescuer's position above the casualty's head it is easy to see the movements of the casualty's chest and to listen to the sound of his exhalations to monitor the effectiveness of the AV. If the casualty is laid on the floor, the rescuer will find that the most comfortable position is if he also lies on the floor supporting himself on his elbows.

If the casualty's face is wet, the pocket mask is likely to slide about and be difficult to hold in position. The simple solution is to wipe quickly both the casualty's face and the rescuer's hands. It is worth including a small towel in the oxygen equipment kit for just this purpose.

Figure 150 Using the pocket mask to administer AV

Limitations of Oxygen Administration

Oxygen should not be considered a treatment for diving emergencies. It is a supplement to normal diving first aid procedures to increase the efficacy of those procedures so that casualties can benefit in terms of a speedier or more complete recovery. No amount of apparent improvement of the casualty's condition should be allowed to detract from the necessity that they subsequently obtain specialist medical treatment.

The administration of oxygen by amateur divers should be limited to casualties who are known to be divers suffering from a diving-related problem. Outside this limitation there are conditions which it might appear to the layman would benefit from the administration of oxygen but for which, in reality, it would be prejudicial.

Sufferers from chronic lung disease often appear cyanosed and to the layman may appear in need of oxygen. In fact, such people have adjusted to a high arterial partial pressure of carbon dioxide and, unlike in normal people, their respiration is controlled by changes in the level of their arterial partial pressure of oxygen. If the oxygen content of their inspired gas is increased only slightly their respiration may cease altogether. In such cases a diver could, with all the best intentions, put the casualty at considerable risk.

Blindness and brain damage can be caused to young children who are exposed to 100 per cent oxygen for even short periods. It can also aggravate the condition of people suffering from high blood pressure by further raising their blood pressure.

Such people who are likely to be adversely affected by oxygen are by the very nature of their conditions, unlikely to be divers. Amateur divers should therefore not administer oxygen to a non-diver unless specifically directed to do so by a doctor.

Many ambulances and rescue vehicles carry cylinders of nitrous oxide, sometimes referred to as 'Entonox'. This is administered to accident casualties as a painkiller. Under no circumstances should this be administered to a diver suffering from decompression illness or burst lung. Any bubbles formed in the blood stream as a result of the condition would not contain nitrous oxide. The appearance of nitrous oxide in the blood would cause an increase in the size of any bubbles as the gas diffused into them until a balance was established. The casualty's condition would therefore be aggravated and his subsequent recovery prejudiced.

AV using a Pocket Mask

Use of the pocket mask will allow AV to be administered in situations when mouth-mouth or mouth-nose resuscitation may not be acceptable due to corrosive chemicals, facial injuries or infectious disease. The use of the pocket mask will also allow oxygen-enriched air to be blown into the casualty's lungs if a suitable supply of oxygen is available. The most effective method of use is to tilt the head using the index, middle and ring fingers of both hands behind the angles of the jaw in front of the earlobes. The lower jaw is pulled upwards so that the teeth of the lower jaw are in front of those of the upper

Figure 151 Casualty breathing from a demand mask

Figure 152 Entonox must not be used on divers suffering from decompression illness

jaw; at the same time the lower lip is pushed downwards slightly using the thumbs and the forehead restrained using the palms of the hands. This technique will open the airway at the same time as opening the mouth and tilting the head (see Figure 153).

Once this has been done the pocket mask can be fitted into place, the lower rim being applied first between the lower lip and the chin. This will help to maintain the retraction of the lower lip. The mask can then be applied to the face and clamped into position using the thumbs to secure the correct position. This also allows the extension to be maintained since it allows the index, middle and ring fingers to operate against the thumb. The rescuer can then apply his mouth to the mouthpiece of the mask and then inflate the casualty's lungs. The use of the mask in the administration of oxygen is explained in the section on oxygen equipment.

Supplementing Oxygen Supplies

Because of space constraints the amount of oxygen available in an emergency is often limited and consequently becomes quickly depleted. If cylinders containing Nitrox are available a secondary means of supplying a casualty with an oxygen enriched breathing supply is possible by using the Nitrox from the divers cylinders. Even with the higher oxygen content Nitrox mixes (such as EAN 50) this will not be anything like as effective as the administration of 100% oxygen (which should always take priority) but will be far more beneficial than breathing air once the oxygen supply has become exhausted.

For all the same reasons as for oxygen, delivery of Nitrox to the casualty should be by an oro-nasal mask. Breathing through the normal diving regulator may in any event be precluded because nausea may render the casualty unable to tolerate breathing through a mouthpiece. A noseclip would also need to be worn to prevent the oxygen concentration being diluted by air inhaled through the nose. This would be even less acceptable to a nauseous casualty.

Oxygen administration equipment should therefore include adapters either to allow the oxygen equipment to be connected to Nitrox cylinders or to enable an oro-nasal mask to be connected to a diving regulator.

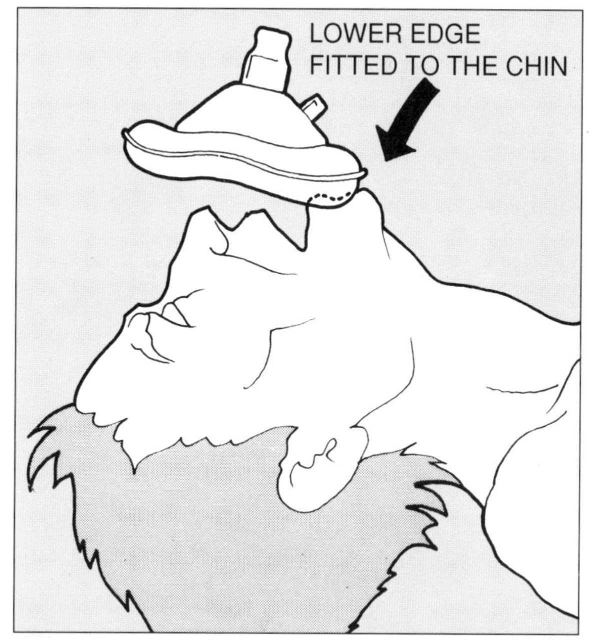

Figure 154

Figure 153 Use your fingers to tilt casualty's chin

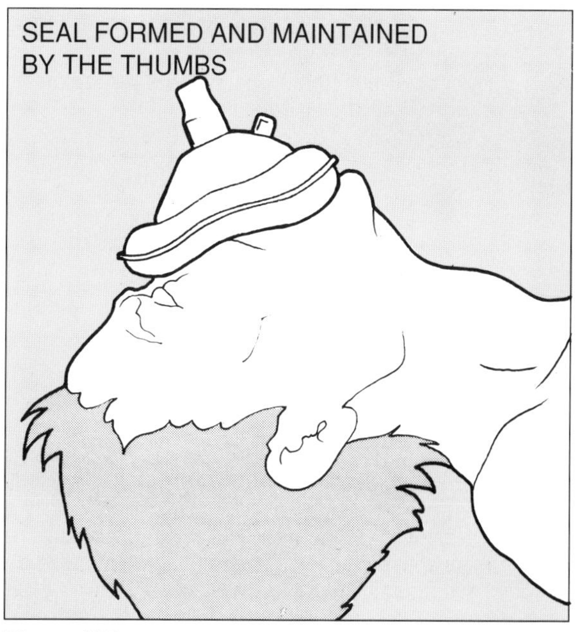

Figure 155

Distress Signals

If an emergency occurs, or is about to occur, then help can best be summoned if there are clearly understood signals available, which indicate the need for urgent action. It follows that the casualty must be aware of these signals, as well as the potential rescuers. Since it is not always possible to determine who will be able to offer assistance at any time or in any location it is essential that the signals to be used should be universally applied and understood. For this reason, avoid the temptation of inventing your own signals to indicate emergency. This would be as useful as inventing a new word for 'Help' or 'Fire'.

Diver-to-Diver

The international signal for distress while underwater is the clenched fist waved from side to side across the chest. It is important that when the attention of the buddy has been gained then a further indication of the cause of the distress is given.

It is also understood that the lack of a response to signals such as 'Are you OK?' should be taken as an indication of a possible emergency.

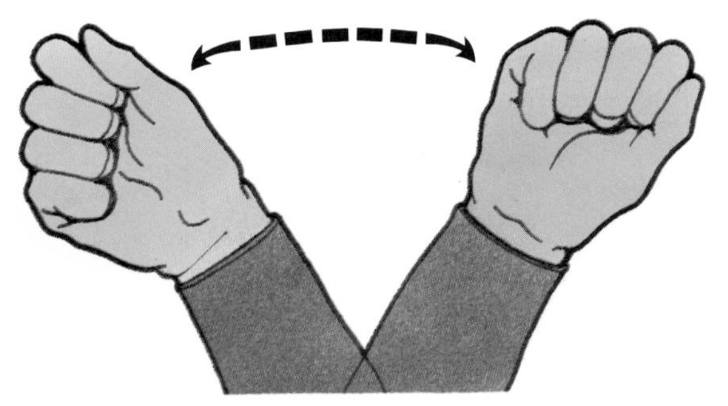

Figure 156 *Distress* A signal, which elicits immediate action to rescue the diver giving it

Diver on the Surface

A diver in trouble on the surface may be signalling either to his buddy, to the diving boat, or to a shore party. In all cases the signal must be interpreted as a 'come and get me' and rescuers should reach the diver in the shortest possible time. Do not wait for a repeat of the signal as it will often be the case that the diver in trouble is no longer able to make the signal. Here again, the lack of a recognizable signal, or of a response to a given signal, should be taken as a need for assistance. If a distressed diver is spotted on the surface then it is important that a member of the surface party maintains continuous visual contact to allow for the possibility of the diver sinking below the surface. This will allow the rescuers to search in the last seen position.

The international signal is an outstretched arm waved from side to side, or indeed any variation from the steady outstretched OK signal

Diver-carried Flags

Since one of the probable causes of distress at the surface may be loss of contact with the surface party or boat, a possible solution is a signal flag, which can be extended on a mast to allow visibility over a greater distance. This will only be practical if it can be collapsed into a sufficiently compact bundle to be carried by the divers during the dive, or if it forms part of a surface marker buoy.

Figure 157 *OK at surface* On surfacing, and if all is well, this signal must be given and maintained until it is acknowledged by the surface cover

Distress at surface – come and get me This signal demands immediate action to assist the distressed diver. Remember also the lack of a signal could mean distress

Figure 158 Diver using flag and mast

Figure 159 Diver holding orange smoke signal

Diver-carried Flares

When diving in conditions where separation from the boat is considered a real possibility thought should be given to the carrying of signal flares or smoke. The items selected should, of course, be totally submersible and waterproof and recommended for use by divers. Divers who have surfaced out of sight of their boat, possibly due to unforeseen tidal streams or a breakdown of the boat, should first establish that they have no other means of contacting their boat or shore party. As soon as they have determined that this is not possible they should employ the appropriate pyrotechnic signals. Remember that in daylight orange smoke is the first choice as it offers the best means of allowing rescuers to pinpoint your position. Red hand-held flares are the second choice, but these are less likely to be seen by people who are not actually searching for you. Use only these colours as they are the internationally recognized distress signals and others may be misinterpreted by other water-users and sailors.

See page 247 of BSAC *Sport Diving* for further details of recognized distress signals.

A diver on the surface who has lost contact with his boat should bear in mind that it is very likely that the distance between him and his boat is increasing all the time. He should therefore be prepared to seek and accept help from any possible quarter. It would be a foolish diver indeed who turned down the offer of help from a passing boat on the grounds that his boat 'should be along any minute now'.

If you are diving at night, or if it becomes night before you reach your boat or the shore, then a powerful lamp or a submersible flashing strobe signal are strongly recommended. A hand flare or rocket might indicate your distress, but would give little assistance to a search party trying to locate you on the surface of an area of sea.

Figure 160 Emergency strobe unit

The Rescue

On seeing any of the distress signals listed above the first duty of the rescuer is to fix the position of the casualty to the best of his ability. Usually this will mean only a direction line, but could be a compass bearing from the current position. Try to maintain visual contact right through to the moment of contact. Bear in mind that if the diver is indicating his position by smoke or a flare then this will soon expire and he may not have another. In the case of smoke go to windward of the smoke cloud. In the case of rockets also bear in mind the influence of wind on the rocket once airborne.

If you are in radio contact with another boat or party who have seen the same signals then immediately attempt to get a triangular fix on the position. This will greatly influence the success of the search.

Emergency Breathing Systems

One of the most serious problems, which a diver can have underwater, is running out of air, or not being able to use the air in his cylinder due to a regulator malfunction. Should this occur, there will be very little time in which to consider the options available, so the divers reaction must be completely automatic. This will only be true if the situation has been allowed for in his training, his choice of equipment, and his buddy briefing.

It may be either yourself, or your buddy, who is out of air. In either case it is important that your actions follow a laid-down pattern that you are both familiar with and can rely on. The actions will depend most of all on the type of equipment being worn by the divers concerned.

Emergency breathing systems can take the following forms:

1. Octopus rig.
2. Second regulator.
3. Pony cylinder.
4. Suitable BC.
5. Redundant breathing system.
6. Combined inflator/regulator.

It will become an essential part of your pre-dive equipment and buddy-check to see which, if any, of the above types of equipment your buddy is using, and to ensure that he has noted the emergency equipment you are carrying.

Octopus rig

The octopus rig got its name from the appearance of a regulator when fitted with two second stages, an HP gauge, and perhaps two inflators. The specific item we are interested in is an extra second stage, additional to that used by the diver.

The intention is that in a 'buddy-out-of-air' situation, the diver will retain his own mouthpiece and offer the octopus to the buddy in need. In practice, the buddy and other members of the diving party will usually be advised to take the octopus directly if they require it, rather than occupy time with an exchange of signals. Once breathing has been re-established the buddy will ensure that the donor is aware that the octopus is in use.

In order to allow the buddy to adopt a position facing, or sometimes alongside, the donor the hose will need to be longer than standard, usually 1 metre in length. Ideally it will be easily identifiable as an octopus, perhaps by tape on the hose, a coloured hose-guard, or a specially coloured front.

Once the buddy has started to use the octopus the donor should wait until both divers have adopted a regular breathing pattern, and then indicate an ascent. The most comfortable breathing pattern will be for the two divers to alternate their inhalation and exhalation patterns so that they are not both inhaling at the same time. Some regulators would be hard-pressed to meet this demand if the divers were at depth. The two divers ascend together at a controlled and steady rate and stay together until they arrive at the surface. The buddy diver should resist any desire to break off just before reaching the surface.

The divers must also bear in mind that the air supply will be consumed at twice the normal rate, since both divers are drawing from the donor's cylinder, and there is a risk of the air supply being depleted if it is not carefully monitored. If the 'out-of-air' situation occurred near the end of a dive it is possible that the donor's remaining air supply would be barely sufficient for a safe ascent, particularly if decompression stops were called for.

An octopus rig will not provide a solution in the event of the failure of the regulator first stage. Should a diver's second stage freeze then the octopus will usually still function normally, although it too might freeze up for the same reason as the primary regulator.

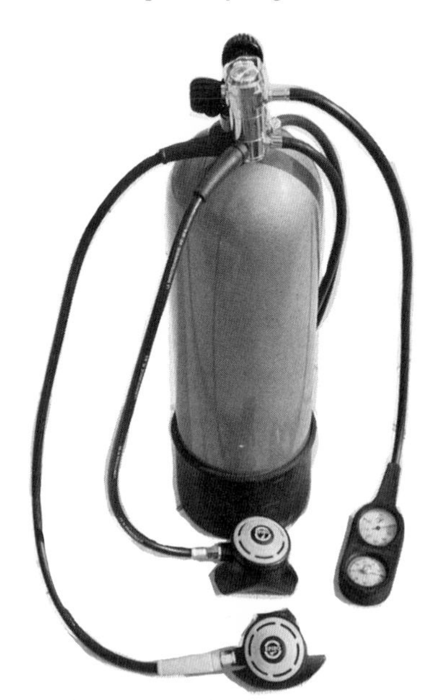

Figure 161 An octopus rig

Second regulator

To carry a second regulator on his system the diver must have a cylinder set which has outlets for two separate regulators. This is the case with many twin sets, which have two outlets at different positions on the manifold, and with some large cylinders which may be fitted with a single-cylinder valve, which splits into two outlets. In an ideal system. both regulators would be of similar type and quality, but in practice we often see a 'spare' regulator which is less efficient and reliable used as the second regulator. An emergency system is of little value if one cannot place full confidence in it in an emergency.

The difference between this configuration and the octopus is that its use by a buddy will have almost no effect on the donor, other than depleting his air supply.

Figure 162 Divers using octopus rig

Apart from this, it would be used in exactly the same way as an octopus, bearing in mind that it will not usually have the longer hose of the octopus. It has the additional advantage that the diver carrying two regulators truly has a secondary regulator, which can be used in the case of any kind of failure of the primary regulator.

Ultimate independence takes the form of a twin set in which each cylinder is equipped with its own regulator. However, if the diver first empties one cylinder completely, before going over to the second, then he will have no emergency regulator available while he is using the second cylinder. The correct procedure in this case is to breathe the first cylinder down to a predetermined pressure, say 30 per cent of the working pressure, then go on to the second cylinder, knowing when this is empty there is still a reserve in the first cylinder.

Pony cylinder

A pony cylinder is a small cylinder which 'rides' on the divers main cylinder. In principle it is an entirely separate supply of air which the diver can use in an emergency, or in a predetermined situation such as decompression. It will be fitted with a second regulator and will therefore be used in exactly the same way as the second regulator described above.

Pony cylinders are usually about 4 to 6 litres water capacity, giving from 30 per cent to 50 per cent additional air to that carried in the divers main cylinder, assuming this to be a 10–12-litre cylinder. It will usually be adequate to ensure a safe return to the surface, and this should be a factor in determining the size of pony cylinder to use. It will need to be securely attached to the diver's main cylinder, with the regulator stowed so that it can be reached either by the diver or by a buddy in need. In some configurations it is attached to the main cylinder in such a way that a buddy in need could pull it away and use it independently.

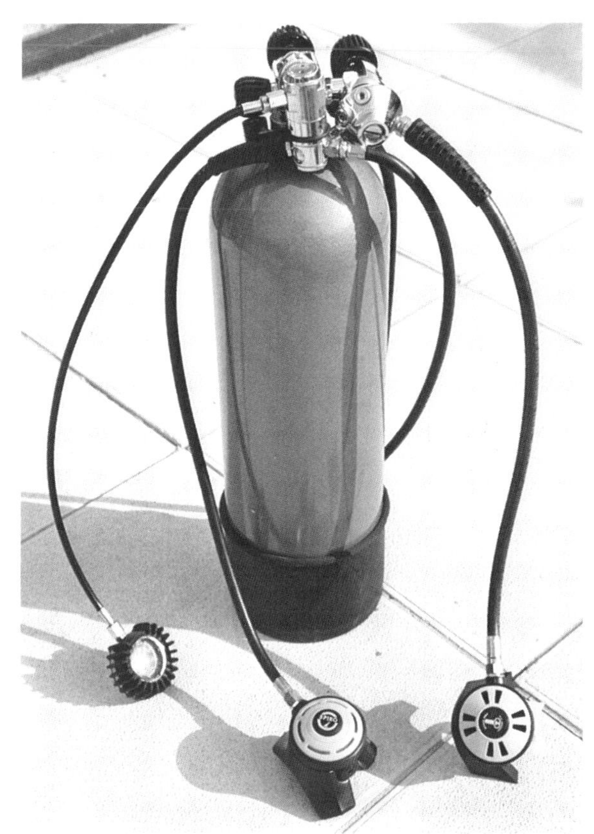
Figure 163 Cylinder with two outlets and two regulators

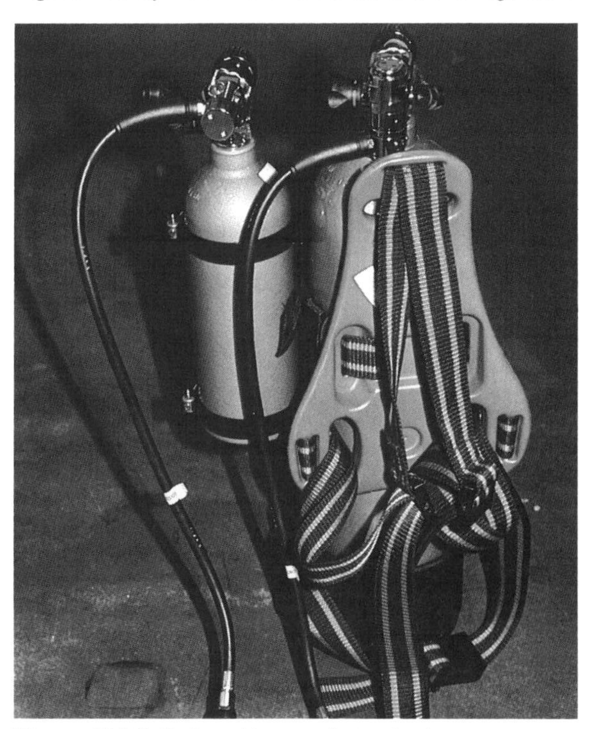
Figure 164 Cylinder with "pony" attached

Suitable Buoyancy Compensator

Many BCs currently in use are fitted with a mouthpiece, which allows the diver to breathe air from the bag of the BC. The mouthpiece is fitted with a diaphragm, which opens a valve when the diver inhales. Provided there is air in the BC he will then be able to inhale it. Exhaled air is exhausted through the mouthpiece.

Air is introduced into the BC manually, either by the diver opening the valve of the BC's air cylinder, or by operating the medium-pressure inflator. As soon as air is introduced into the bag the diver's buoyancy increases and he will probably start to ascend. If the diver then inhales this air it is transferred from the BCs bag to the diver's lungs, but maintains its buoyancy until the diver exhales. With practice a competent diver could continue to breathe in this fashion until the air supply is exhausted. However, it is a complex procedure and one, which it is considered, would be unlikely to be successful in most emergencies. Its practice was therefore dropped from diver training some time ago, the emphasis being moved to the correct use of octopus and secondary regulator systems.

In some situations it is considered that a rescuer could pass their primary regulator to a casualty and continue the ascent breathing from the BC. The complexities of this procedure are not believed to offer a reliable safety procedure.

Redundant breathing system

A recent introduction has been a system totally dedicated to emergency breathing, not forming any part of the diver's normal diving equipment. This equipment consists of a small cylinder of compressed air equipped with its own special regulator. It is carried by the diver and in an 'out-of-air' situation can be used by the diver, or handed to a buddy in need.

The air supply is usually very limited, sufficient for only a few minutes, and therefore suitable only for a direct ascent. Its use will be most successful if the diver using it is familiar with its use and has had the opportunity to practise with it previously.

Combined inflator/regulator

This item combines the function of the drysuit, or BC inflator with that of an octopus or secondary regulator. It is located on the BC or drysuit and is used in the normal way as an inflator, being connected to the suit or BC by a corrugated hose, and to the regulator by a conventional MP hose.

In an 'out-of-air' emergency the principle to be followed is that the donor hands the casualty his primary regulator, with its standard-length hose, and uses the combined regulator for his own breathing. This is the most comfortable procedure as the combined regulator has a relatively short hose and would unnecessarily

Figure 165a Monitoring the ascent during a controlled buoyant lift

Figure 165b Keeping the casualty above the rescuer helps to control the ascent

restrict the casualty's movement. It will, of course, also serve the diver in the case of a failure in his primary second-stage regulator.

Associated with this type of equipment are hybrid arrangements, which make use of a single MP hose from the regulator first stage to feed an emergency second stage as well as the BC inflator. In practice these fall between the 'combined' regulator and the octopus in their application depending on the length of the hose fitted. Some of these types reserve the octopus for the casualty, while others will demand that the donor uses the emergency regulator while passing his own to the casualty.

In general, it is considered sound emergency practice to avoid the need to give up any part of your own life-support system if it can be avoided. A well-planned diving system should provide for a secure main system, and an emergency system, which will not deprive you of your own air supply just when you need it most.

Some of the systems listed serve principally as self-rescue equipment, some are buddy-rescue-only equipment, and some can have both functions. It follows that divers who are taking some responsibility for the other members of the diving group will want to equip themselves to be able to assist the others. Divers not carrying such a direct responsibility, other than that implicit in the 'buddy', system may prefer to equip themselves principally for self-rescue. In practice, the out-of-air situation is avoidable and careful monitoring of your high-pressure gauge, and your buddy's is the best precaution you can take.

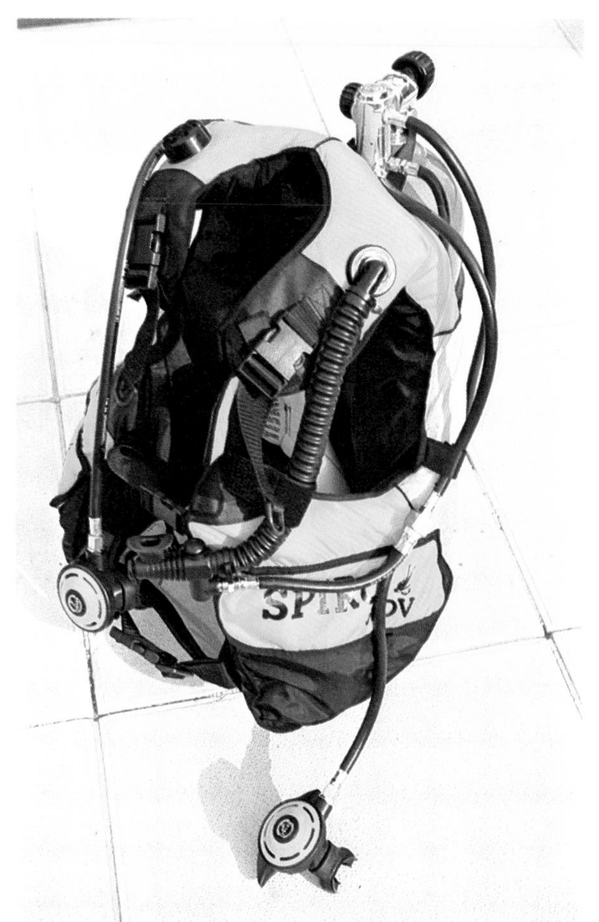

Figure 167 A combined octopus/inflator system

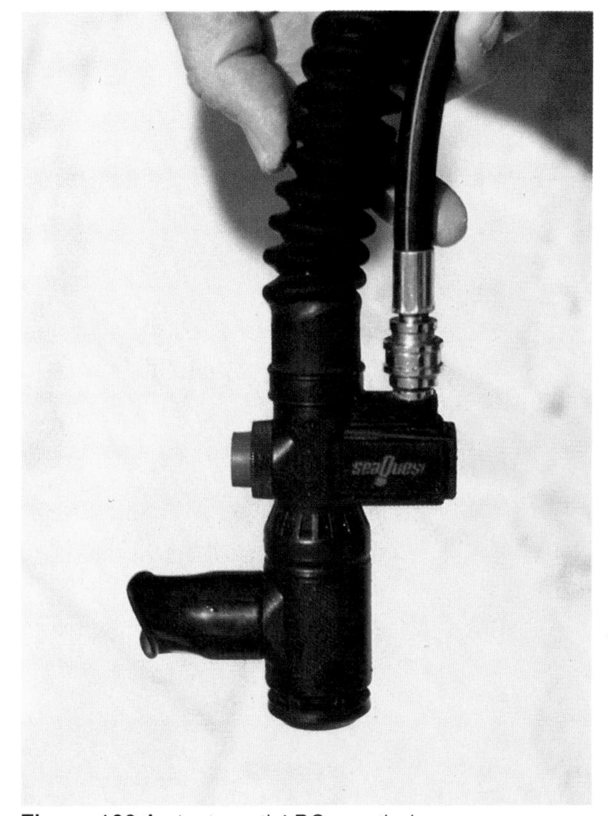

Figure 166 An 'automatic' BC mouthpiece

EQUIPMENT	**SELF-RESCUE**	**BUDDY-RESCUE**
OCTOPUS		*
SECOND REGULATOR	*	*
PONY CYLINDER	*	*
SUITABLE BC	*	?
REDUNDANT SYSTEM	*	*
COMBINED INFL/REG.	*	?

Throwing Aids

The priorities of an emergency situation may indicate that it is best to throw something to the casualty (see Pages 50–51) so consider what will be most useful.

You will be throwing because it is considered inappropriate, or not yet appropriate to wade, swim or row toward the casualty. In many cases others may be doing these things already, but throwing will give additional or earlier help.

Fixed or Loose

The first consideration is whether to throw a loose object, or something with a line attached. Almost always it will be more useful to select something with significant intrinsic buoyancy, to which a line can be attached. In some cases almost any buoyant object will bring benefit provided it is thrown soon enough. The choice will depend on the state of the casualty.

Buoyant Conscious Casualty

The problem foreseen in this case is that the casualty is unable to return to the boat or shore by their own means, although they are in no danger of sinking. They need a line thrown quickly and accurately that they can catch and secure and which the rescuer can then haul in until the casualty can regain the boat or the shore. A buoyant object on the end of the line, or the use of floating line, will avoid the possibility of the line landing near the casualty but sinking before he can secure it.

The line must be long enough to stretch from the rescuer to the casualty, bearing in mind any increase in the distance resulting from tidal streams or wind drift. It is particularly distressing to throw an object only to find that the line will not reach.

Non-buoyant Conscious Casualty

The first priority is to give the casualty anything, which will support him or increase his buoyancy so that time can be won for the second priority of retrieving him. In a diving location you will find many different types of buoyant objects. On a diving boat there will often be fenders, line floats, shot line buoys, divers' BCs, as well as conventional life-buoys or even life-rafts. Any of these thrown as accurately as possible towards the casualty may assist and may also prove useful subsequently to the rescuers who may be swimming or wading to the casualty.

Any of these objects will prove more valuable if a line can be attached to them, once more of the right length. Should it turn out that the line is not sufficiently long then in this case, if the object is considered likely to reach the casualty if the line is cast loose, and the casualty will benefit from the buoyancy of the object, then it may be best to let it go.

It is possible that the person to whom you are throwing a line is a rescuer who has reached the casualty but needs the assistance of a line to regain the boat or shore. In this case you can expect due co-operation, but you will

Figure 168 A selection of buoyant objects suitable for throwing

Figure 169a

Figure 169b Throwing a bagged line

be hauling in two people, rather than one and this may require you to seek assistance from other members of the party.

Throwing a Buoyant Object

By their very nature some of these buoyant objects are difficult and unwieldy to throw and may have too little weight, or too great a surface area, to reach the required distance. If an object such as the popular 25-litre container is selected as a float, then it will throw better if it is partially filled with water, although this will also reduce its final buoyancy.

Throwing a large object into the wind will usually prove unsuccessful, while its performance downwind may prove quite satisfactory.

Throwing a Line

If you have the opportunity during an emergency to select your line then look for a properly coiled floating line. Hold the coil in your left hand so that it is free to pay out forward and take the top four or five strands in your right. Throw the rope in your right hand as accurately as you can with an underhand swing preceded by one or two back and fore swings to gain some momentum and to gauge direction. Be sure that you secure your end of the line so that it cannot be pulled from your hand by the effect of wind, tidal stream, or the casualty. If the only line available is in an uncoordinated pile then it will probably be worth a swift attempt at coiling it before attempting to throw it. An uncoiled line is most unlikely to pay out as desired, or to reach its full length.

Prepared Throwing Lines

It is possible to prepare lines specifically for throwing. On larger boats and ships a lighter line with a weighted end (a monkey's fist knot) is used and the heavier hauling or mooring warp is tied on to this. Such a throwing line would be of great use in an emergency.

Specially designed throwing lines are also available which are made to allow the rescuer to throw more accurately and over a greater distance. The line is attached to a bag into which it is fed and in which it is stored. The loose end of the line is finished with a loop, which the rescuer places around his right wrist. The bag has a tape loop attached, which the rescuer uses as a handle. It is swung back and forth as with a rope and then thrown underhand toward the casualty. The rope pays out of the bag as it flies through the air so that the main body of the rope is thrown and lends its weight and momentum to improve the accuracy of the throw. In practice these ropes are very useful, but it is important that they are located where they might be needed so that they can be employed quickly in an emergency.

A rescue quoit – a weighted plastic ring with a light line attached is sometimes found fitted in small boats and also increases throwing range

Figure 170a Throwing a coiled line

Figure 170b

Figure 170c

Recompression Therapy

Recompression therapy is the definitive treatment for decompression illness as well as being used to treat a number of medical conditions, which may be unrelated to diving (e.g., carbon monoxide poisoning and gas gangrene).

The symptoms of serious decompression illness are caused by inert gas bubbles (mainly nitrogen in the case of air divers) which collect in the venous circulation when pressure is reduced during ascent. Some of these bubbles may cross into the arterial circulation to produce further damage. Air emboli (bubbles of air) may find their way directly into the arterial circulation as a result of damage to lung tissue. In both conditions blood flow to important tissues is blocked by bubbles in blood vessels causing tissue hypoxia.

Recompression reduces the size of these bubbles in two ways. The increased ambient pressure reduces bubble volume and forces nitrogen back into solution. In addition, an increase in the partial pressure of oxygen in the inspired gas increases the oxygen supply to the hypoxic tissues.

No other treatment of decompression illness is as effective or as safe as recompression therapy. Delays in initiation of recompression therapy after dysbaric illness reduce the chance of complete recovery. So no other treatment or first aid measure should be allowed to delay transfer to a recompression facility. In cases where there is doubt whether symptoms are due to dysbaric illness or a coincidental medical condition, it is recommended that recompression be undertaken to see whether symptoms resolve. Such a 'trial of compression' often causes rapid resolution of symptoms and is unlikely to harm the casualty.

The recompression chamber consists of a large steel cylinder in which the ambient pressure can be increased. The pressure inside the chamber is normally stated as the identical depth in sea water, which would result in the same ambient pressure. Thus, when the ambient pressure inside the chamber is raised to a pressure of 2 bar, we talk about being at a depth of 10m (although the chamber does not actually go underwater).

The basic approaches of recompression therapy have been described in the *Sport Diving* Manual. In theory, there are a large number of possible treatment profiles. The depths at which treatments are given, the rates of descent and ascent, and the gas mixtures breathed can be varied infinitely. The recompression profiles used are designed not only to eliminate the bubbles which originally caused symptoms, but also to allow return to the surface at a rate which ensures that gases are eliminated from the body without producing further problems. In practice, a sport (scuba) diver developing dysbaric illness is likely to receive one, or a combination, of the following types of recompression profile.

1.Shallow Oxygen Treatment

Compression for a period of time to 18m (60ft). Whilst at this depth there are repeated twenty-minute periods breathing oxygen via a mask followed by five minutes breathing air from the chamber atmosphere. These air breaks are used to reduce the risk of lung oxygen toxicity. 100 per cent oxygen is not usually used at depths greater than 18m because of the possibility of fits due to neurological oxygen toxicity. After treatment at 18m the chamber pressure is slowly reduced (over thirty minutes) to 9m, where further oxygen treatment with air breaks is given before slow decompression to atmospheric pressure. Royal Navy Tables 61 and 62 and US Navy Tables 5 and 6 use this technique.

2.Deep Air Treatment

Compression to 50m (165ft), during which time air is breathed. It is usual to interrupt the ascent to atmospheric pressure with stops at 18m and 9m. Once above 18m, provided it is available, the gas breathed is oxygen with air breaks (as for shallow oxygen treatment). R N Tables 4 and 6a use this technique. (Oxygen is available at all UK chambers, but not necessarily abroad.)

3.Special Gas Mixture Treatment

Mixtures of oxygen and helium (oxy-helium) have been used in Britain for treatment of serious decompression illness, which has not responded to shallow oxygen, or deep-air treatment. The casualty and tender are left at the dive depth or deeper for several hours in order to give adequate time for improvement. Decompression takes place over the next four to five days, whilst the casualty is given hyperbaric oxygen treatment. This technique avoids the problems of nitrogen narcosis in the casualty or tender at depths greater than 50m, but it is very expensive. In other countries, different gas mixtures may be employed (e.g., oxygen-rich mixtures, more than 20 per cent but less than 100 per cent oxygen, at depths between 18 and 50m).

Amateur diving is invariably conducted at sites remote from recompression facilities, so that in contradistinction to professional and much armed forces dysbaric illness, there is a delay between the amateur diver developing symptoms and reaching a recompression chamber. The length of this delay may affect the treatment given. At the chamber the casualty will be assessed before recompression. A diving medical officer, who may not be personally present at the chamber, will direct the treatment used. Below are outlined some commonly used guidelines, but the actual choice of treatment table will be affected by the clinical judgement of the diving medical officer, the response to treatment, and the facilities available.

In most mild cases, e.g., a skin bend, the casualty may merely be observed and not be recompressed unless symptoms worsen. If the casualty has a limb bend he will usually be taken to 18m on an oxygen table. If the pain disappears rapidly at this depth, he will remain there for about forty-five minutes (RN Table 61, US Navy Table 5) before being brought to the surface. If the pain does not resolve within ten minutes at 18m, a longer period at this depth will be required. (RN Table 62, US Navy Table 6 stipulates at least four and three-quarter hours in the chamber.)

A long oxygen table at 18m is also used for the initial treatment of decompression illness, involving the cardiorespiratory or neurological systems. It could be that the casualty has only noticed pain in a limb but when examined before entering the chamber a neurological abnormality was found, which will be assumed to be the result of decompression illness.

If signs or symptoms get worse, or fail to improve during the initial 20-30 minutes of shallow oxygen treatment, it will often be necessary to take the casualty down to 50m on air. A deep air table is commonly used from the outset when symptoms are definitely or even possibly due to air embolism, or if oxygen is not available for shallow treatment for decompression illness. If the USN Table 4 is used the casualty will be in the chamber for over 38 hours. Oxy-helium can be used to take divers deeper than 50m if symptoms fail to resolve after a period at 50m on air.

Oxy-helium is most often required for dysbaric illnesses arising after dives to depths greater than 50m. Unfortunately, a large number of recompression chambers are not equipped with helium. As a result, divers who develop dysbaric illness after dives to depths greater than 50m may not receive optimum treatment. Divers should be aware that if they develop neurological symptoms as a result of such a dive, then the possibility of complete resolution of their symptoms is diminished.

A multiplace chamber allows the casualty to be accompanied by a qualified tender and, if necessary, a Doctor. The tender makes sure that the chamber, its air locks, and other equipment are functioning correctly. He will examine the casualty periodically to assess the response to treatment. He administers oxygen, first aid and additional therapy (discussed below) to the casualty. Use of a multiplace chamber also allows medical or surgical procedures, such as insertion of a chest drain, if the casualty has a pneumothorax as well as air embolism.

Occasionally, more than one casualty is treated simultaneously in a multiplace chamber. When this happens, it is necessary to pressurise the chamber to treat the symptoms of the most severely affected casualty.

Sometimes it is necessary to treat a casualty in a single place chamber. Most of these are situated in hospitals, where they are used for treating non-diving routine medical conditions, e.g., gas gangrene and radiotherapy of tumours. Most of these chambers can operate at pressures up to 3 bar (20m) and can therefore be used for administering oxygen therapy at 18m. However, in these chambers the patient breathes the chamber atmosphere, so that the chamber must be filled with oxygen and this does not allow for air breaks. As a result, treatment tables are considerably modified. These chambers do not allow ready access for medical staff to the casualty. They should not be used for casualties who are unconscious, or likely to become unconscious, because of the risk of cessation of breathing.

Occasionally, portable single place chambers are used for transporting casualties to a multiplace chamber. This should only be done after consultation with the multiplace chamber operators, since the single-place chamber may be incapable of locking onto or passing through the door of the multiplace chamber. As a result, a transfer under pressure may prove impossible.

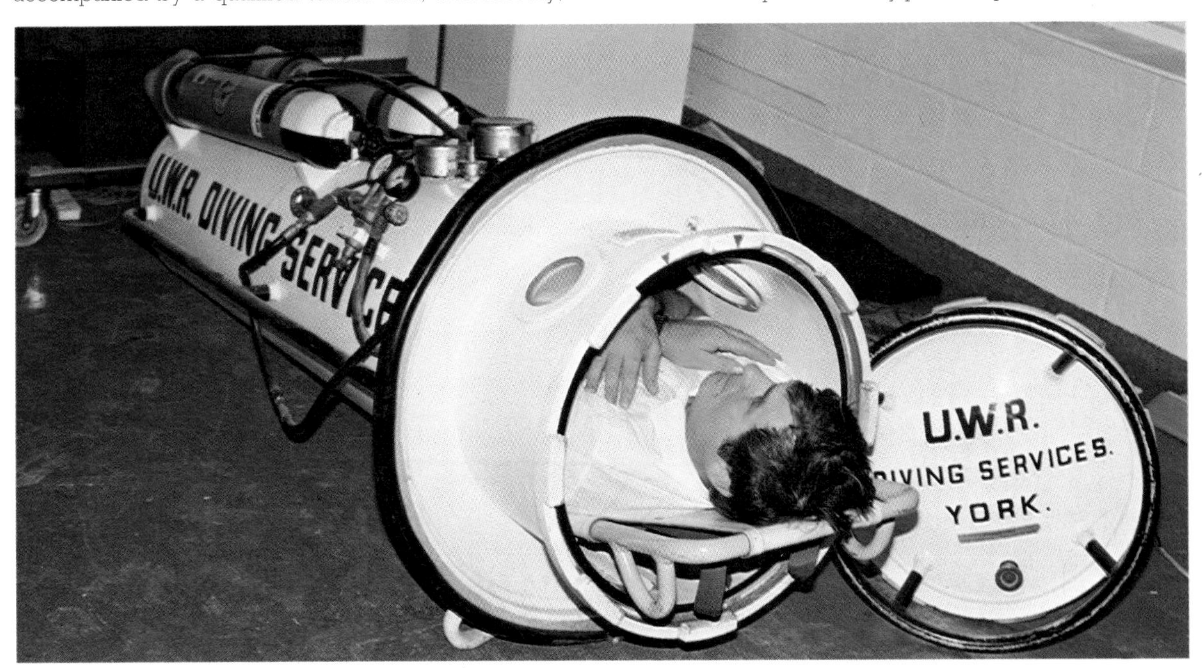

Figure 171 A single man recompression chamber

Additional Treatment

The supervising doctor selects the additional treatment given to a casualty with dysbaric illness.
Some of those commonly used are described below.

1.Fluids

During decompression illness, fluid passes out of the blood vessels. This causes the same effects as dehydration including, when severe, a reduction in blood pressure. In addition, during treatment the heat in the chamber may cause further fluid loss. Also, blood cells stick to each other and to the gas bubbles. These effects lower tissue blood flow and exacerbate hypoxia. This can be partially reversed by giving fluids. Fully conscious patients may be given fluids by mouth, unless there is a possibility of vomiting. Intravenous fluids may be considered for those with cardiorespiratory or neurological symptoms, or when consciousness is impaired.

2.Urinary Catheter

When spinal decompression or unconsciousness prevents the casualty passing urine a urinary catheter will be used so that fluid balance can be observed.

3. Drugs

Steroids may be given to reverse inflammation and swelling in the brain and spinal cord. Sedatives are given to treat fits, and may be used for sickness and dizziness. Drugs, which reduce the clotting ability of the blood, may be used in some circumstances to reduce the stickiness of blood cells.

Post-treatment Considerations

If symptoms persist after recompression therapy, further courses of shallow oxygen treatment may be given on a daily basis. This may produce some improvement in symptoms, but will not necessarily produce complete resolution.

After completion of recompression therapy the patient should not ascend to altitude, or fly in an aircraft, for up to seventy-two hours, depending upon which treatment table has been used. The supervising medical officer will advise on the appropriate interval.

A diver who has required treatment for dysbaric illness is not permitted to resume diving again with the British Sub-Aqua Club unless certified fit to do so by a BSAC Medical Referee.

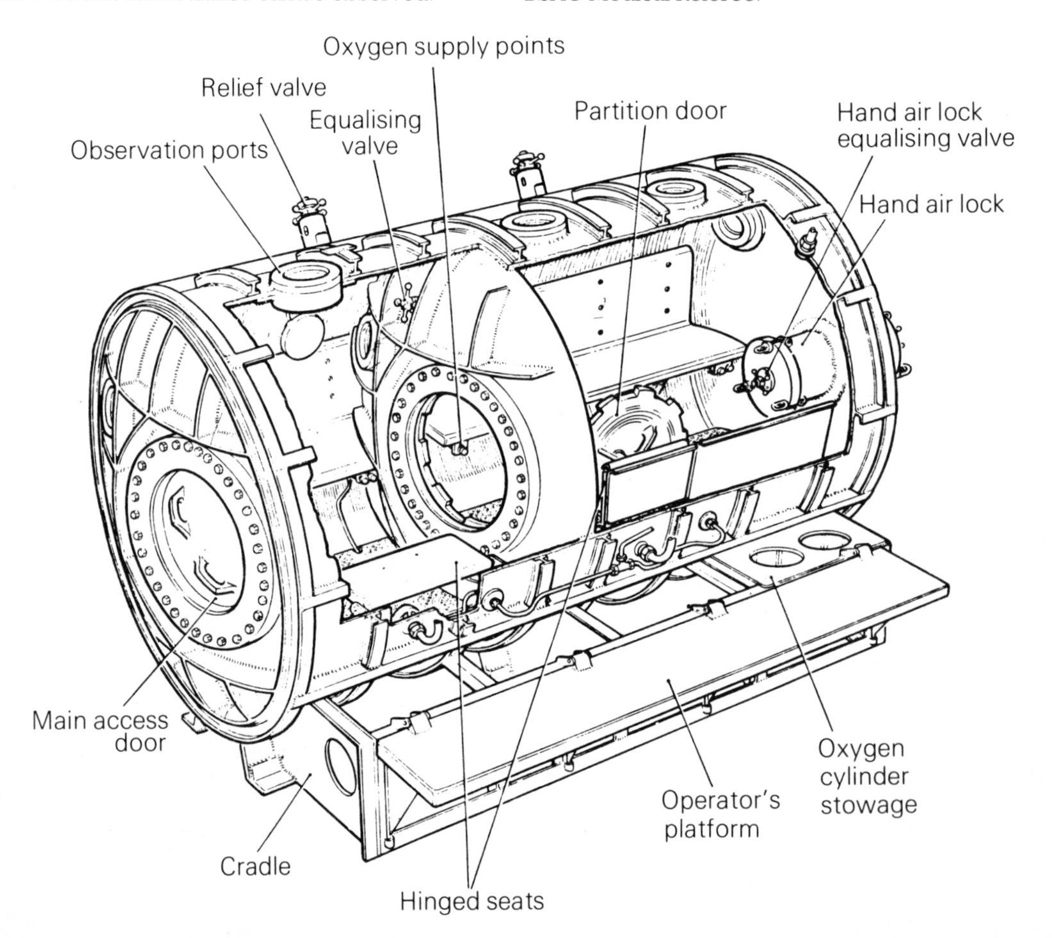

Figure 172 Layout of a recompression chamber

Figure 173 A Recompression chamber

Figure 174

Hazardous Marine Life

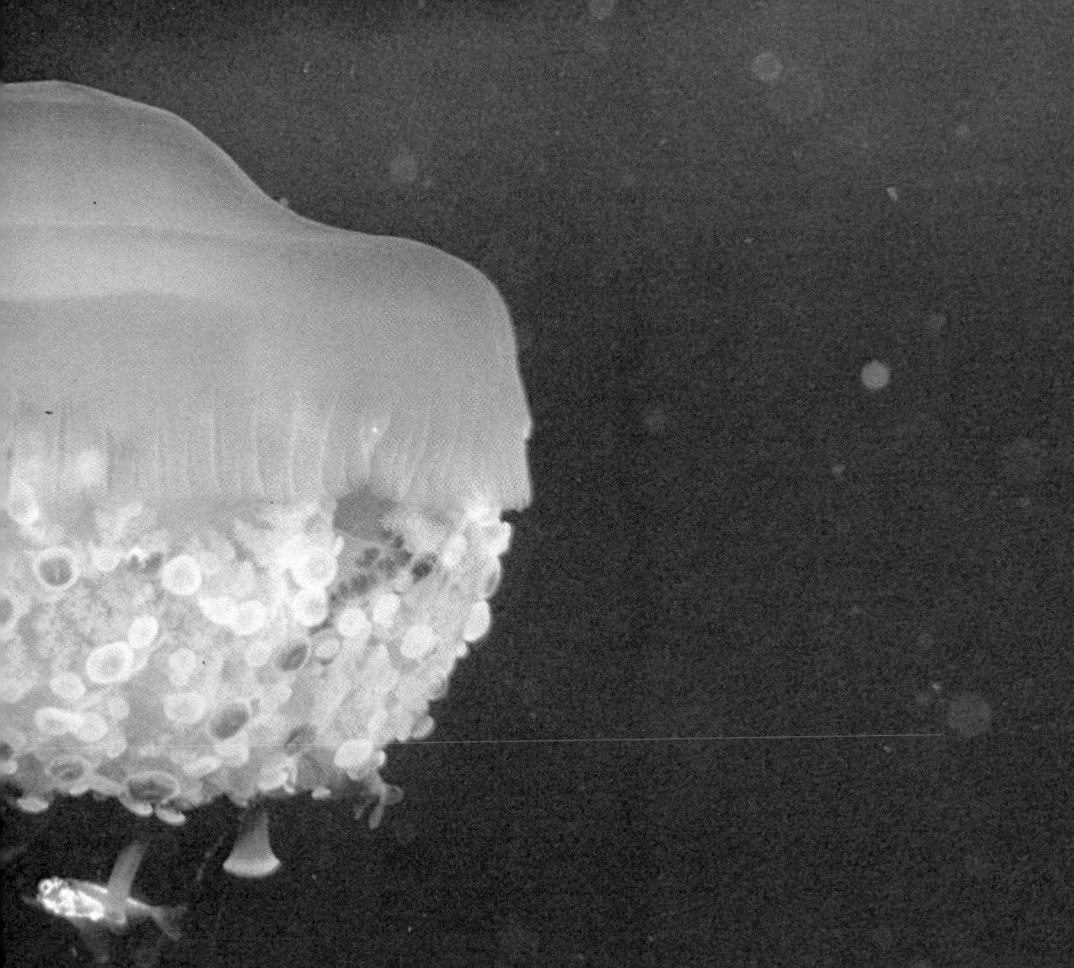

Dangerous Marine Creatures

This section is not intended to be a comprehensive guide on the treatment of all bites and stings which divers may suffer. It concentrates on the main creatures to be aware of, as far as physical injury is concerned, but many of the first aid measures will be appropriate for creatures, not mentioned below, which cause damage to divers and swimmers.

We can divide dangerous creatures into two sections, those, which bite, and those which sting. Secondary infection is always a risk, especially with bites, and divers in remote locations need to administer effective first aid if permanent disability or even death is to be prevented.

Divers in warm tropical waters are more likely to have larger areas of bare flesh exposed and are therefore more susceptible to attack by marine creatures, and as there are more dangerous animals in such waters, the risk of such attack must be that much greater. Added danger results in the case of infections which often occur after the initial bite or sting. The severe shock, which follows a serious bite or sting, is also something which those carrying out first aid need to be aware of and should include in their treatment.

There are very few marine creatures in the waters surrounding the British Isles and Europe, which are considered to be harmful to divers. Those which can cause damage rarely do so unless provoked. There are, however, creatures, which need to be treated with respect, and if bitten or stung by such creatures, divers need to be aware of the appropriate first aid.

Figure 175 Angler fish

Figure 176 Conger eel

Bites

One or two fish have been known to inflict nasty wounds when they have been threatened. The Angler fish *(Lophius piscatorius)* is a common and grotesque creature found around the British Isles and grows up to 2m in length. They have a huge mouth with rows of sharp teeth, which are capable of inflicting very nasty wounds (see Figure 175). Once they have hold of a casualty they will not let go until they are dead, and anyone bitten could suffer severe wounds, which will carry the risk of secondary infection.

Essential treatment will include the use of pressure pads to control the bleeding and the cleansing of the wound, preferably with fresh water with added disinfectant. The casualty of such attacks will need further hospital treatment to include a tetanus injection if such protection has lapsed.

Conger Eels have been known to bite divers, but only when they have been disturbed by hands being poked inadvertently into their lairs. They can cause a nasty wound, however, and treatment is the same as for bites by Angler fish, mentioned above.

Sharks

Sharks are rare around the British Isles, with only one or two species being in evidence, and those being mainly concentrated in the English Channel, off the coasts of Devon and Cornwall. Although they have the equipment to inflict a nasty bite they have virtually never been recorded as doing so and cannot be considered a threat.

Shark attack in other parts of the world is fortunately very rare and the reputation of sharks is fiercer than the reality. There are, however, a handful of dangerous species which do cause severe injury and death to swimmers and divers each year. Probably the best known is the Great White or White Pointer. This shark grows up to several metres in length and often attacks in very shallow water. In Australia, swimmers in water of standing depth have been fatally attacked, and divers engaged in spear fishing activities are especially at risk. A close watch also needs to be kept for Tiger Sharks and Hammerheads. Most shark casualties have no warning of the attack and some Australians maintain that another reason to dive with a buddy is that it reduces one's chance of being attacked by 50 per cent!

Shark attack wounds are usually severe and extensive. Injuries range from teeth puncture marks to tears and amputation of limbs. The rough texture of a shark's skin also leads to skin abrasions. Massive blood loss is common with severe shock. Casualties display low blood pressure, cyanosis and a rapid pulse. Blood poisoning, tetanus and gangrene are real threats to those who survive an attack.

Experts in the treatment of shark attack maintain that the casualty *should not be moved* until their condition has stabilized, and that medical aid should be brought to them. Under no circumstances should the casualty be rushed off to hospital as this reduces the chances of survival.

Immediate first aid must be to stop bleeding, preferably by the use of pressure pads, broad tourniquets if necessary or finger pressure on the arteries. Towels or items of diving equipment may be used to improvise. If medical or paramedical assistance is on hand then painkillers such as morphine should be administered. If possible, the administration of a blood transfusion or plasma will help stabilize the casualty's condition.

Drugs given by mouth must be avoided, as the patient will probably require surgery and this action could preclude the use of an anaesthetic.

Figure 177 Hammerhead shark

Crocodiles

Crocodiles cause severe wounds and treatment is similar to that for shark attack. The Estuarine or Salt Water Crocodile is quite common in northern Australia, Indonesia, Malaysia and other parts of Southeast Asia and can grow to 6m, while the Nile Crocodile is a hazard in many parts of equatorial Africa.

Figure 178 A Nile crocodile

Figure 179 Blue-ringed octopus

Blue Ringed Octopus

The Blue Ringed Octopus, which inhabits the shoreline of large parts of Australia, is often found in rock pools and is a small, attractive creature, its body being covered with blue or purple rings. Children are often at risk as they search for shells in rock pools, as are divers who find them and handle them whilst on a dive. The bite is generally painless and may go unnoticed until the toxic effect of the poison takes hold. Within minutes paralysis begins and leads invariably to a cessation of breathing. The frightening thing is that although conscious and aware of all that is going on, the casualty is unable to communicate with rescuers, who may even think he is dead! Other symptoms may include difficulty with vision and swallowing and progressive numbness around the head and mouth. The bite itself becomes swollen and red

Treatment must include artificial ventilation, which may need to continue for several hours. A ligature above the wound is effective treatment, if applied properly, and venom should be removed from the wound.

Sea Snakes

Sea snakes are common creatures in many tropical regions of the world and come in many species. Their venom is the most toxic of any snake, affecting the central nervous system with paralysis of limbs and the breathing system. Weakness and muscular contractions are common, and in serious cases it is important to treat the casualty with the appropriate antivenin as soon as possible. Antivenins, however, have inherent dangers in their use and medical assistance must be obtained before they are administered. Although all sea snakes are highly venomous they are not usually aggressive and approach the diver with curiosity rather than malice. Sudden, panicky movements should be avoided, however.

First aid for a sea snake bite will include the application of a ligature, if the injury is on a limb, and the removal of any venom on the surface of the skin. CPR may be necessary in serious cases.

Moray Eels

Moray Eels have a reputation, which exceeds reality, and because of their fierce appearance have been blamed for all manner of sins. They are to be avoided, however, if threatened and a diver is most at risk when poking hands into holes without checking what is there. Certain types of Moray Eel are said to be poisonous, but it is likely that secondary infection from a wound, giving similar symptoms to poisoning, is the real cause of so-called venomous bites. Morays have sharp teeth, however, and deep wounds will result from a bite. Control of bleeding is essential first aid, as is thorough cleansing of the wound with appropriate disinfectant.

Figure 180 Sea snake

Figure 181 Moray eel

Stings

European Jellyfish

There are few creatures to worry about in British waters with regard to stings. The Common Jellyfish (*Aurelia aurita*) poses no threat.

More sinister, however, are the stings from another quite common jellyfish, *Cyanea lamarckii*, which are blue or brown and white in colour, up to 50cm in diameter and trail numerous tentacles, often many metres in length. They are often found in large numbers around our coast, particularly the English Channel and North Sea, and can inflict very painful stings. Their tentacles are to be seen trailing from buoy lines, which are often, left marking sites of wrecks. Divers not wearing gloves, who are in the habit of pulling themselves down such lines often find out to their cost the danger of doing this. Several divers each year are also stung around the exposed parts of the face while descending buoy/anchor lines. Severe pain and numbness are the symptoms of such stings, which fortunately only last for a short time. Relief can be gained by the application of vinegar or methylated spirit over the affected area, or from proprietary painkillers.

Figure 182 Compass jellyfish

Box Jellyfish (Sea Wasp)

The Box Jellyfish or Sea Wasp (*Chironex fleckeri*), which inhabits much of the Queensland and northern coasts of Australia, is among the most dangerous marine creatures to be encountered. The long tentacles are almost invisible in the water, and at certain times of the year they drift in large numbers close to the beaches, where they can create havoc. When the tentacles make contact with the skin, a highly toxic venom is discharged. Death can occur within minutes as a result of the venom or the drowning which follows the casualty's inability to help himself. Pain is excruciating and the tentacles leave skin lesions, which can lead to permanent scars if the casualty survives.

The casualty must be quickly removed from the water and the tentacles removed. Rescuers must take great care not to handle the tentacles, as they themselves will be badly stung. Post-attack shock is usually severe so relief from the stings is an important first aid step. The application of methylated spirits or vinegar is a tried and tested method of giving relief. Any alcoholic spirit is a suitable alternative to the above. Tentacles should be removed by scraping with a suitable object. Some anaesthetic ointments are reasonably effective, but if medical or paramedical assistance is available then morphine or a similar drug should be administered intravenously. Look out for impaired breathing and circulation, and CPR may be necessary. Fortunately an antitoxin has been developed in Australia and its application as soon as possible will greatly increase the chances of survival. A vaccine to give immunization is currently being developed.

Prevention is always better than cure, however, and many divers in the waters where the Box Jellyfish occurs now wear suits made of a thin Lycra material, which gives good skin protection, the water being too warm for wetsuits. They are effectively a body stocking. Nylon tights or stockings are another alternative giving some degree of protection.

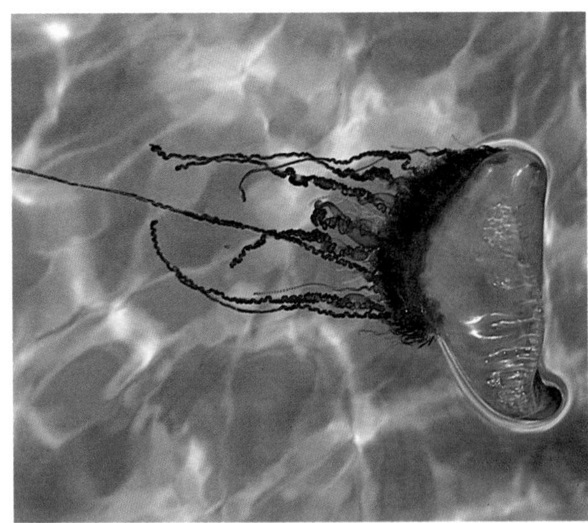

Figure 183 Portuguese man o' war

Stonefish

Stonefish are more likely to be a source of danger whilst walking over shallow coral reef areas than when diving, although divers must be very careful where they place their hands, as the Stonefish is extremely well camouflaged. Extremely toxic venom is injected through the dorsal fins, which are erected when the fish feels threatened. Severe localized pain is experienced and developing symptoms may include shock, coma, respiratory failure and possibly death. Stonefish antivenin is really a last resort as it can have dangerous reactions and *must* be administered by a doctor. Relief can be gained by application of local anaesthetics. If these are unavailable, relief can be gained by submerging the limb in hot water (45°C) for about half an hour. This has the effect of relieving the pain and can help in destroying the poison. CPR may be necessary in serious cases.

Figure 184 Stonefish

Scorpionfish, Lionfish, Weevers

Several varieties of the above types of fish have a fairly worldwide distribution in warm waters. They all inject poison through their dorsal fins, in a similar manner to the Stonefish. Symptoms are painful but not generally as potentially lethal as the latter. The first aid treatment is similar to that for a Stonefish.

Figure 185 Lionfish

Figure 186 Cone shell

Venomous Cone Shells

There are more than 400 species of venomous Cone Shell. Divers who are unaware may think they are picking up an attractive shell. The Cone Shell is extremely venomous and stings through a 'poison dart' attached to a tube, which strikes from within the shell. Symptoms are similar to that of a sea snake bite, as is the treatment. Cone Shells are capable of stinging through clothing, so beware.

Figure 187 Crown of thorns starfish

Crown of Thorns Starfish

The spines of the Crown of Thorns starfish are venomous and often break off when the starfish is stood on. Redness and numbness occur in the casualty, with possible paralysis as well as nausea and vomiting. Bathing the wound in hot water (50°C) is effective first aid and, if spines are embedded, local anaesthetics may be required to cleanse the wound. Steroid creams and antibiotics may be required for further treatment.

Figure 188 Sea urchin

Sea Urchins
Tropical Sea Urchins are a ubiquitous hazard to divers and to those walking over rocky areas. The long-spined, black Diadema, with spines, which easily penetrate through shoes, are especially hazardous. Some urchins have venomous spines while others are more of a danger because of the risk of infection from spines trapped in the wound. Spines must be removed, but they have the habit of breaking off when this is attempted. Bathing in hot water (50°C) gives relief from the pain and also softens the skin to allow removal of spines. It is important that all spines are removed, as secondary infections almost certainly will occur if not.

Fire coral
Fire coral, which grows in a variety of shapes, has stinging cells (*nematocysts*) which are long enough to penetrate the human skin and toxic enough to cause a reaction. The sting feels like a burn – hence the name fire coral. The effects can vary from a minor irritating rash, to large, slow-healing sores, which if left untreated may lead to a serious infection. The affected area should be washed with soap and water, cortisone or antihistamine cream applied, and covered with a sterile dressing. Protective gloves should be worn when diving around fire coral.

Figure 189 Fire-coral millepora

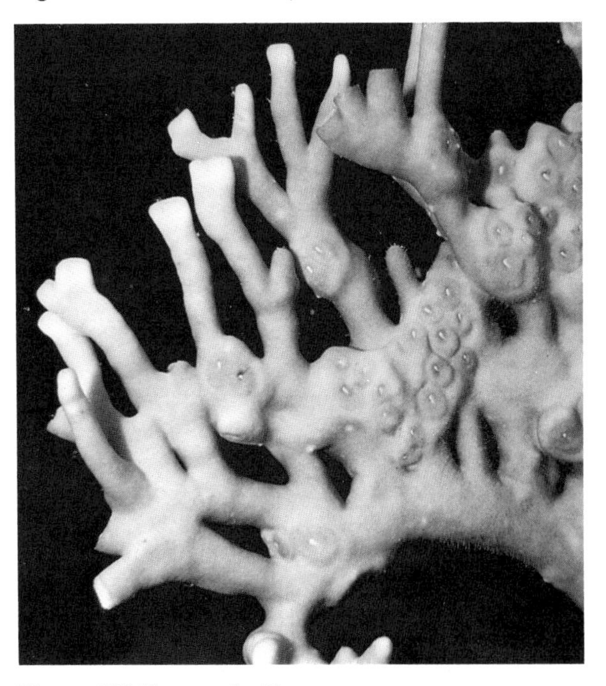

Figure 190 Fire-coral millepora

Appendix 1 – BSAC Rescue Awards

The Diver Rescue Specialist

The Diver Rescue Specialist is a diver recognized as having received comprehensive training in all aspects of sport diver rescue. In addition to the normal rescue skills training inherent in the BSAC training programme, the Diver Rescue Specialist will have participated in additional BSAC courses and awards.

Lifesaver Award
Advanced Lifesaver Award
First Aid for Divers Course
Oxygen Administration Award
Practical Rescue Management Course

He will thus have a well-founded knowledge of, and practical ability at, all aspects of rescue from prevention, through the various elements of a rescue situation, until the casualty is delivered into the care of appropriate medical aid.

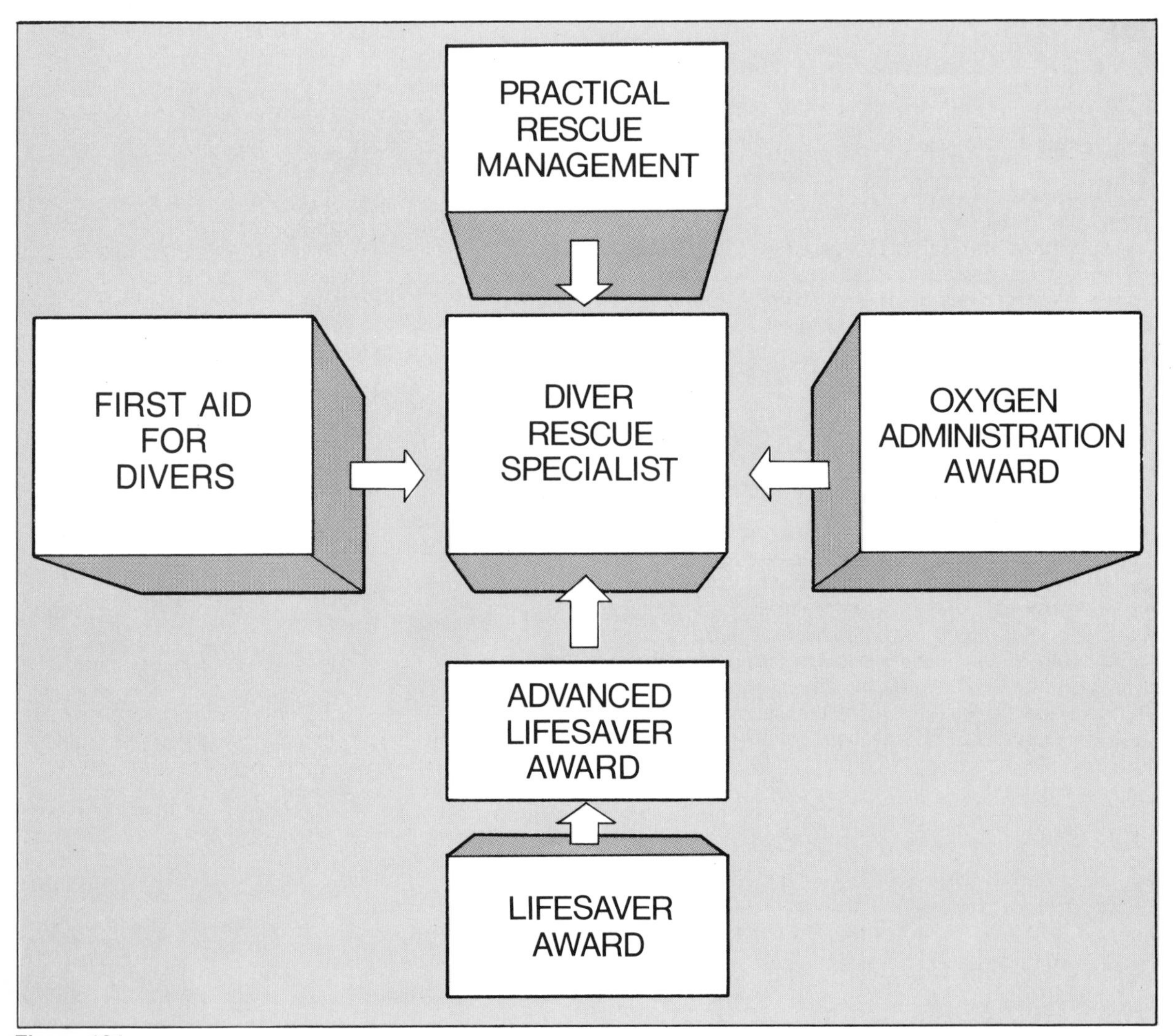

Figure 191

Lifesaver Award

The primary aim of this Award is to examine, under pool or sheltered water conditions, the lifesaving proficiency specifically applicable to divers. A secondary aim, however, is to examine the ability of divers to make use of their lifesaving skills in more general non-diving situations.

The syllabus comprises two sections – a 'dry' theoretical and practical test and a 'wet' practical test.

The 'dry' theoretical test comprises a question and answer session covering the topics of:
respiratory and circulatory anatomy
respiratory and circulatory physiology
lifesaving and relevant first aid

The 'dry' practical test requires the candidate to demonstrate a satisfactory standard of the following skills:
artificial ventilation
diagnosis of cardiac arrest
chest compression combined with artificial ventilation
action for dealing with a casualty who vomits and for putting the casualty in the recovery position.

The 'wet' practical test comprises a number of elements:

Throwing rescues – using both ropes and buoyant aids. These simulate the situation where the rescuer is in a boat with a conscious casualty a short distance away. The rescuer and casualty will be dressed and equipped as though for the real situation.

Snorkelling rescue – this represents the situation where a conscious casualty is too far away to be aided by a throwing rescue. The rescuer, equipped with snorkelling equipment, swims 50 m to the casualty, inflates the casualty's buoyancy aid and then tows the casualty 50 m before assisting the casualty to simulate an exit from deep water into a boat.

Full equipment rescue – with both rescuer and casualty dressed for normal diving. This exercise simulates the rescue of an unconscious, non-breathing casualty from depth to a shelving shore. The rescue commences in 4 m of water. The rescuer lifts the casualty using a controlled lift to the surface, inflates the casualty's buoyancy aid to secure him at the surface and then commences artificial ventilation while towing the casualty to shallow water.

Swimming rescue – this element assesses the rescuer's ability in a non-diving situation. No diving equipment is worn by either casualty or rescuer. The rescuer swims 25 m to an unconscious, non-breathing casualty floating face down in the water. The casualty is towed 10 m to a deep water support and artificial ventilation administered before the casualty is removed from the water. As a second aspect the rescuer also assists an unconscious but breathing casualty to keep his head above water for 2 minutes to simulate supporting a casualty while waiting for other assistance to arrive.

Advanced Lifesaver

The aim of this Award is to assess lifesaving proficiency specifically applicable to divers in realistic open water conditions. As such it builds on the Lifesaver Award by repeating many of the elements but in real, rather than simulated conditions. An element is also introduced to cover first aid for injuries likely to be encountered in diving and diving related activities.

The syllabus comprises four sections:
Throwing exercises
Controlled buoyant rescue from depth
Surface rescue
First aid

For all elements rescuers and casualties are equipped with their normal diving dress and equipment.

Throwing exercises – this element follows the same format as the similar exercise in the Lifesaver Award but with the added complication that the rescuer makes the throws from within a small inflatable boat rather than from the relative stability and clear space of a poolside.

Controlled buoyant rescue from depth – this element simulates the rescue of an unconscious, non-breathing casualty from a depth. The casualty is raised towards the surface using a controlled buoyant lift but, as this is only a simulated rescue, for both decompression considerations and as a demonstration of the control of the ascent the lift is halted at a depth of 6 m. Following a normal ascent to the surface the rescue is resumed by the rescuer inflating the casualty's buoyancy device and then administering artificial ventilation while managing the removal of the casualty from the water into a small boat.

Surface rescue – this element portrays the alternative ending to the previous rescue where there is no boat available and the rescuer must tow the casualty to shore. The exercise commences as though the lift had just been completed and the casualty is towed to shore where the rescuer manages the removal of the casualty from the water.

First Aid – the casualty is considered to be breathing but suffering from some injury likely to have been suffered during the course of diving or diving related activities. The rescuer administers first aid using whatever resources are available on site. This is supplemented by a short theory test to assess a broader spectrum of first aid understanding.

First Aid for Divers

Aim

To instruct divers in actions to be taken when dealing with injuries and illnesses which are likely to occur during diving activities.

Classroom lesson: Principles and Priorities of First Aid
Purpose
First Aiders role and responsibilities
Priorities
Hygiene

Practical lesson: Casualty evaluation
Obvious injuries
Hidden injuries
Levels of consciousness
Priorities
Passing on information

Classroom lesson: Summoning assistance
Diving incident
Non-diving incident

Classroom lesson: Shock
Types
Signs and symptoms
Treatment

Classroom lesson: Wounds and bleeding
Types of wound
Direct pressure
Pressure points
Elevation
Disadvantages of tourniquets
Major external bleeding
Internal bleeding

Practical lesson: Bandaging and pressure points

Classroom lesson: Burns
Types of burn
Blisters
Sunburn
Treatment

Classroom lesson: Miscellaneous injuries and conditions
Minor pressure injuries
Blast injury
Seasickness
Hypothermia
Hyperthermia
Exhaustion

Classroom lesson: Injuries to muscles, bones and joints
Function of the skeleton
Types of fracture
Signs and symptoms
Immobilizing the fracture
Dislocation

Practical lesson: Dressings Splinting and moving casualties

Classroom lesson: First aid kits
Small boat kits
Dive Marshal's kit

Classroom lesson: Open forum
General discussion of subject
Course debriefing
Issue endorsements

The order in which the syllabus is listed represents the recommended sequence for the course. The contents of the course should emphasize practical first aid, but with sufficient explanatory theory to put the practice into its true context. Instructors should base their teaching on the information given in this manual and the BSAC *Sport Diving* Manual.

Figure 192 Lifesaver awards badges

Oxygen Administration Course

Aim
To teach the administration of oxygen as a means of increasing the effectiveness of the first aid treatment of diving accidents. (It is implicit in this definition that the treatment takes place on the surface, either in a boat or on dry land. There is no question of administering oxygen to a submerged diver.)

Classroom lesson: *What is oxygen?*
Role of oxygen in respiration
Presence in the atmosphere
Non-flammability

Classroom lesson: *Diving accidents*
Conditions and symptoms
Benefits of using oxygen

Practical lesson: *Revision of AV/CPR*
Solo rescuer – AV
Two rescuers – CPR

Practical lesson: *Use of the pocket mask*
Aid to AV

Classroom lesson: *Oxygen administration equipment*
Cylinders
Regulators
Masks
Storage
Oxygen supplies

Classroom lesson: *Oxygen administration in practice*

Practical lesson: *Use of oxygen equipment*
Assembly
Administration – breathing casualty
– non-breathing casualty
– with cardiac arrest

Practical lesson: *Casualty evaluation*
Examining the casualty
Recording information

Assessment Test

The order in which the syllabus is listed represents the recommended sequence for the course. Instructors should base their teaching on the information given in this manual and the BSAC *Sport Diving* Manual.

Practical Rescue Management

Aim
To teach divers the management of rescue activities and the rescue skills involved.

Classroom lesson: *The rescue process*
Prevention
Anticipation
Priorities

Classroom activity: *Revision of CPR*
Revision and practice using resuscitation manikins

Classroom lesson: *Rescue management part 1*
Recovering the casualty
First Aid priorities
Casualty records
Summoning the emergency services

Practical lesson: *Rescue management scenarios*
Practical exercises involving different types of casualty and different situations

Classroom lesson: *Rescue management part 2*
Lessons arising from practical scenarios
Dealing with relatives / friends
BSAC Incident Reporting Scheme
Dealing with media enquiries

Classroom lesson: *Open forum*
General discussion of subject
Course debriefing
Issue endorsements

Appendix 2 – BSAC Medical

Diving is a risk sport, which involves physical exertion and exposure to unusual physical stresses, e.g., weightlessness, cold and altered partial pressure of gases. To dive safely the diver must have sufficient physical, medical and mental fitness and must have obtained suitable skills through training to cope with all possible conditions that he may encounter on the dive. The decision about whether someone may dive safely includes assessment of his ability to cope with a dive and also assessment of how his lack of fitness or skills may affect his buddy. Questions of whether the diver is likely to require rescue by his buddy or, if the roles are reversed, whether he will be able to rescue an incapacitated buddy must be considered.

It is clear that adequate levels of skills and fitness required for a dive in one set of conditions might be inadequate for a dive in different conditions. It is the Branch Diving Officer (or his representative) who has final responsibility for deciding whether an individual has sufficient skills and levels of fitness to undertake a particular dive. This decision must be based on recommendations laid down in BSAC literature, including *Safe Diving Practices* and BSAC manuals.

Medical fitness, certified by a doctor, is one of the requirements for an individual to dive with the BSAC. Branch Diving Officers have a responsibility to ensure that members comply with this requirement.

A thorough medical examination will usually detect medical conditions or diseases likely to produce problems underwater. However, such examinations are conducted on land and may not detect psychological problems, which are likely to produce adverse reactions or diving stresses. Instructors have a responsibility to look for and overcome adverse psychological stresses in their students.

Similarly, medical examinations do not test for physical fitness. Branch Diving Officers must assess their members to decide whether they possess sufficient physical fitness, as well as skills, for a particular dive. The Officer must be satisfied that a diver has a valid medical certificate to dive with the BSAC. Recommendations about the requirements and frequency of examinations for medical fitness are advised by the National Medical Committee to the National Diving Committee and Council.

Currently, members require a medical examination on entry. Because diseases likely to affect diving become more common as people age, repeat medicals are required at progressively shorter intervals. If a diver is under thirty years of age his medical certificate will be valid if the examination was conducted within the previous five years, if he is aged thirty to fifty years his medical examination should have been within the previous three years; and after the age of fifty annual medicals are required. A repeat medical examination is also required before resumption of diving after all serious illnesses or major surgery. A chest X-ray is only considered necessary if recommended by the examining doctor.

Any qualified medical practitioner (including all general practitioners and hospital doctors) may conduct these medical examinations. In cases where it is difficult to make a decision about medical fitness, the diver may be referred to one of the nationwide panel of medical referees, who have special interest in diving medicine and who may arrange special investigations.

A diver who has suffered from serious decompression illness (requiring recompression therapy), pulmonary barotrauma, or other diving-related illness is considered unfit to dive unless approved to recommence diving by one of the BSAC's medical referees.

The medical standards of the BSAC are outlined on the BSAC Medical Certificate for the guidance of the examining doctor. The basic principles behind them are fairly simple. A medical condition is deemed to disqualify a person from diving if:

a) it is likely to cause incapacity or unconsciousness in, or under, the water, so as to place at risk the health or life of the diver or his buddy;
b) it is considered to predispose to serious diving-related illness, whether these are apparent in the water or sometime later;
c) it is likely to make diving-related illness more difficult to diagnose or treat.

An example of a condition, which might cause incapacity or unconsciousness in the water, is epilepsy. A fit underwater is likely to result in drowning and also places the diver's buddy at risk during a rescue attempt. In such circumstances the buddy has to execute a rapid ascent and might suffer pulmonary barotrauma as a result.

Lung diseases and obesity are conditions likely to predispose to diving-related illnesses. Some lung diseases result in difficulty in expelling air from the lungs, which can result in burst lung during a normal ascent. This produces symptoms during the ascent or soon afterwards, often whilst still in the water. The propensity of nitrogen to dissolve in fatty tissues and these tissues' poor blood supply, means that excessive body fat predisposes to decompression illness, which usually occurs after leaving the water.

A number of conditions, which can cause unconsciousness or neurological abnormalities e.g., multiple sclerosis, can be difficult to distinguish from pulmonary barotrauma and decompression illness. As a result, the patient could receive inappropriate or even harmful treatment.

Variations to the disqualifying conditions may be made for disabled divers on an individual basis by BSAC medical referees or the National Medical Committee in consultation with the National Diving Officer, Regional Coach or NDC Adviser on Diving and Disability. For example, a person who is paraplegic (paralysed in both legs as a result of a spinal disease) might not be able to dive in an unrestricted manner. The reasons for this are that he might be considered to be less capable of rescuing his buddy than a diver without such a disability, and also because there is evidence that those with pre-existing spinal disease are more predisposed to spinal decompression illness than other divers, and that when it occurs in them, decompression illness is more difficult to treat. As a result, such an individual might be allowed to dive as a disabled diver but with certain qualifications. He might be advised to dive only in a trio with experienced divers (so that he would not be required to take part in a rescue) and instructed to dive shallow and well within the no-stop time limit, so that there would be little risk of developing decompression illness. It might also be necessary to insist on annual medicals, irrespective of the age of the diver.

Below are mentioned some of the medical conditions which are of particular importance to divers. This is not a complete list.

Ear, Nose and Throat

The ears and sinuses contain air. They are greatly affected by pressure changes. If the narrow passages connecting the middle ear or sinuses to the nose are obstructed because of inflammation or scarring, the diver may be unable to equalize pressure in these structures. This will result in pain in the ear or face. If nothing is done to equalize the pressure, bleeding into the sinus or perforation of the eardrum could occur. Perforation of an eardrum underwater allows cold water to enter the middle ear, which results in vertigo and disorientation. Wax blocking the outer ear also produces pressure damage, including perforation of the eardrum and reversed ear. Wax might also prevent water draining from the ears after diving and result in ear infection. It is therefore wise to have the ears checked for excess wax each year. Although long-standing sinus problems may interfere with diving, others find that their chronic sinus problems improve when diving.

Cavities in teeth and beneath loose fillings contain air. Pain can occur as a result of volume changes during a dive. Loose-fitting dentures can be dislodged during a dive, particularly during air sharing and can obstruct the air passages.

Respiratory System

Active lung disease such as tuberculosis, absolutely precludes diving. This is because it will impair gas transfer and hence exercise performance. It also predisposes to burst lung.

Pulmonary barotrauma is most likely to occur in those lung conditions in which narrowing of the air passages occur. This reduces the rate at which the individual can breathe out. As a result, during an ascent, gas volume may increase more rapidly than the diver can breathe out and produce lung damage. Because the greatest volume changes occur nearest the surface, the risk of pulmonary barotrauma is greatest in the last 10m of the ascent. Consequently, individuals with lung disease should never be given a restricted medical certificate for shallow diving. Narrowing in air passages may occur in a small part of one lung (e.g., cyst or tuberculous cavity) or throughout both lungs (e.g., chronic bronchitis or asthma).

Asthma tends to affect young and otherwise fit individuals. Most of the time the air passages are of normal size, but the bronchioles narrow considerably during an asthma attack. This narrowing may be produced by allergens such as pollens and furs or by cold, emotional stress or exercise. Attacks of allergic asthma do not seem to occur in the clean air of the coast, or when breathing purified air supplied by a diving cylinder. Therefore, individuals with only allergic asthma may be allowed to dive provided tests by a medical referee are satisfactory. Obviously, those who get asthma attacks during emotional stress, cold exposure or during exercise are not permitted to dive. They will be exposed to all these conditions when diving.

It should be remembered that anyone with a cold or chest infection has inflammation and narrowing of their bronchioles. This could result in burst lung, so they should not dive.

Smokers, even when they do not have a chest infection, often have narrowed air passages and seem more likely to get pulmonary barotraumas than non-smokers.

Anyone who has had lung trauma or surgery or a burst lung, whether the result of diving or not, will require special consideration before they can dive again.

Heart and Circulation

Serious diseases of the heart and circulation, including high blood pressure, may reduce exercise capacity or may produce unconsciousness or heart failure, when the sufferer is exposed to weightlessness, cold and stress. These conditions disqualify the individual from diving.

Nervous System

A history of epilepsy disqualifies an individual from diving, because of the risk of unconsciousness underwater. Other neurological diseases may predispose to decompression illness and may complicate its treatment. A medical referee must decide in each case. A past history of mental illness does not necessarily disqualify although current treatment usually does, partly because the drugs used to treat these illnesses can have adverse effects in divers.

Temporary Unfitness

Temporary unfitness to dive results from any infection of the ears, sinuses or respiratory system, as well as from any major illness. Sedative drugs, including alcohol, can persist in the body for over twenty-four hours. Alcohol consumption should be considered a reason for temporary unfitness.

There is believed to be an increased probability of birth abnormalities in babies born to women who have dived deep in pregnancy, especially in the first three months. Therefore, it is recommended that pregnant women should limit their diving. The BSAC recommends that pregnant women should not dive below 20m and should stay no longer than the no-stop time for that depth minus five minutes.

Index

Illustration Acknowledgements

Thanks are due to the following for allowing the use of copyright photographs:

Kurt Amsler, figures 181, 187
Dan Burton, figures 1, 6, 41, 54, 59, 60, 112, 113, 124, 125
Mike Busuttili, figures 10, 11, 12, 23, 46, 47, 64, 73, 77, 78, 80, 83-86, 88, 90, 96, 97, 100, 101, 104, 107, 110, 111, 115, 140, 145, 148, 150, 151, 158, 160, 161-170, 173
Adrian Clarke, figures 128, 146, 147
Leo Collier, figure 180
Dave Crockford, figure 138
Andrew Mounter, figure 183
Christian Petron, figures 174, 176, 190
Linda Pitkin, figure 188
Chris Prior, figure 186
Carl Roessler, figures 177, 184, 185
Peter Scoones, figure 189
Jonathon Scott, figure 178
Dave Shaw, figures 3,4, 5, 114, 129, 133, 134, 136, 159, 171
Roy Waller, figure 182
Alan Wilkes, figure 144
Barry Winfield, figures 130, 131, 139
Dave Wybrow, figure 132

All the artwork for this book was specially commissioned from Rico Oldfield